CANNABINOIDS

CANNABINOIDS

LINDA A. PARKER

The MIT Press | Cambridge, Massachusetts | London, England

The MIT Press
Massachusetts Institute of Technology
77 Massachusetts Avenue
Cambridge, MA 02139
mitpress.mit.edu

The MIT Press would like to thank the anonymous peer reviewers who
provided comments on drafts of this book. The generous work of academic
experts is essential for establishing the authority and quality of our
publications. We acknowledge with gratitude the contributions of these
otherwise uncredited readers.

This book was set in Chaparral Pro by New Best-set Typesetters Ltd. Printed
and bound in the United States of America.

Library of Congress Cataloging-in-Publication Data is available.

ISBN: 978-0-262-05139-2

10 9 8 7 6 5 4 3 2 1

EU Authorised Representative: Easy Access System Europe, Mustamäe tee 50,
10621 Tallinn, Estonia | Email: gpsr.requests@easproject.com

CONTENTS

SERIES FOREWORD

The MIT Press Essential Knowledge series offers accessible, concise, beautifully produced pocket-size books on topics of current interest. Written by leading thinkers, the books in this series deliver expert overviews of subjects that range from the cultural and the historical to the scientific and the technical.

In today's era of instant information gratification, we have ready access to opinions, rationalizations, and superficial descriptions. Much harder to come by is the foundational knowledge that informs a principled understanding of the world. Essential Knowledge books fill that need. Synthesizing specialized subject matter for nonspecialists and engaging critical topics through fundamentals, each of these compact volumes offers readers a point of access to complex ideas.

The MIT Press's publishing mission benefits from the generosity of our donors, including Diana Chapman Walsh.

This book is dedicated to the memory of the "father of cannabis research," Professor Raphael (Raphi) Mechoulam. Up until the very end of his life, as throughout his entire life, he was devoted to discovering the functioning of the endocannabinoid system. Raphi brought "big ideas" to the annual conference of the International Cannabinoid Society (ICRS), which he originated in 1990. These "big ideas" often shaped the hypotheses he and his collaborators tested and, most often, verified. These collaborative enterprises unlocked the mysteries of this newly discovered system that appears to regulate almost all bodily functions. Raphi never lost sight of the mystery and instilled this wonderous spirit into so very many researchers and their students around the globe. He is dearly missed.

Here I tell the story of how a few scientists in the last eighty years discovered how a plant that has been used medicinally and recreationally for thousands of years, cannabis, produces its effects on the human body. This journey of discovery has revealed a previously unknown biological system, the endogenous cannabinoid (endocannabinoid) system, that has since been found to be one of the most ubiquitous systems in our brain and body; this system affects almost every aspect of our biology.

The intent of this book is to present the current state of knowledge about cannabinoids (cannabis and endo-cannabinoids). As the debate about the medical use of cannabis persists, it is important to understand the fundamental science and the potential harms/benefits of this drug that is becoming increasingly legally accessible. Such knowledge is particularly pertinent considering the proliferation of strains and extracted tinctures of high potency Δ^9-tetrahydrocannabinol (THC), the intoxicating constituent in cannabis, that are increasingly becoming available in market. In the chapters that follow, the risks and benefits of cannabis use are reviewed with regards to its specific effects on the brain. The most common concerns about potential health risks of cannabis use include (1) the potential for addiction (chapter 4), (2) the potential of cannabis to precipitate psychosis (chapter 5), and (3) the potential of cannabis to impair brain development and learning (chapter 6). Cannabis also has potential medical benefits, including (1) reduction of stress (chapter 5), (2) regulation of appetite (chapter 7), (3) reduction of nausea and vomiting (chapter 8), (4) reduction of pain (chapter 9), and (5) regulation of seizures and other neurodegenerative disorders (chapter 10).

My intention was to focus on the high-quality evidence for potential therapeutic benefits or risks of cannabinoids accrued since the publication of my earlier book, *Cannabinoids and the Brain* (MIT Press, 2017). Considering the lessening restrictions on the public use of cannabis in recent years, I was surprised to discover that such evidence remains elusive, apart from the use of the primary nonpsychoactive constituent in cannabis cannabidiol (CBD) for the treatment of childhood epilepsy. Based on high-quality human randomized control trials (RCTs), CBD (Nabidiolex, GW Pharmaceuticals) was approved in 2018 by the Federal Drug Association (FDA) in the United States for this indication. Although preclinical evidence is revealing a myriad of other potential therapeutic benefits of cannabinoids, it is essential to validate these findings in human patients. Future high-quality RCTs are essential to ensure that anecdotally reported benefits of cannabinoids are not merely placebo effects. Raphi Mechoulam would urgently argue for the need for the basic research findings to find their way into the clinic.

INTRODUCTION TO CANNABIS

An altered state of consciousness, euphoria, relaxation or stress reduction, increased enjoyment of food tastes and aromas, distortion in time perception, increased appreciation of humor and music, joviality, introspection, and increased sensuality and creativity—these are some of the typically reported "psychoactive effects" experienced by cannabis users. Among the several hundred compounds identified in the cannabis plant, it is Δ^9-tetrahydrocannabinol (THC) that produces these psychoactive effects or the "cannabis high." THC is present in most parts of the plant but is most highly concentrated in the droplets of sticky resin produced by glands at the base of the female flowers. Higher doses of THC may produce anxiety and even psychosis, particularly among inexperienced users or even among experienced users in an unfamiliar context.

The primary nonintoxicating cannabinoid that has received considerable recent investigation is cannabidiol (CBD). A review of the journal database program PubMed reveals a dramatically accelerating base of CBD research over the past forty years, with over thirteen hundred research articles on CBD published in 2020 alone.

Neither THC nor CBD are present in the fresh cannabis plant. Instead, the fresh plant contains their precursors, THC acid (THCA) and CBD acid (CBDA), which are slowly decarboxylated (i.e., losing their acidic function) in response to heating (when marijuana is smoked). Unlike THC, THCA does not produce psychoactive effects. In addition to CBD, there are several other cannabinoid components (discussed in chapter 3), including flavonoids and terpenes giving cannabis its distinctive odor and flavor, that have received little study. Human clinical trials will be necessary to harvest the "treasure trove" of cannabinoids found in cannabis. Indeed, the combined effects of these plant cannabinoids produces what has been called an "entourage effect."[1]

Varieties of cannabis plants differ in THC content. Hemp contains less than 0.3 percent THC, but it is high in CBD. In the 1960s, THC content in the most common strain, *Cannabis Sativa*, was about 2–3 percent, but specialized cultivation techniques have increased that content to up to 25 percent today in some varieties. Because THC and CBD are produced by the same precursor in the plant, Cannabigerol (CBG), as the THC content increases in the plant the CBD content decreases. Therefore, a by-product of the selective breeding for high-potency strains of cannabis has been the breeding out of CBD such that very low levels are typical in street cannabis. This suggests that today's cannabis differs considerably from the cannabis that was available years ago, both in its effects on mental health and cognitive functions as are discussed in later chapters.[2]

A Brief History of Cannabis Use

Cannabis use became widespread in Western nations during the revolutionary 1960s, but it has been used throughout recorded human history around the world as a fiber, food, and herbal medicine. The first evidence of cannabis cultivation was over ten thousand years ago in central Asia. Most of the pharmacological properties of cannabis that are only now being scientifically studied were known and many were used in medicine for the treatment of numerous pathologies in ancient times.

The first historical evidence of cannabis use in traditional medicine has been credited to the legendary Emperor Shen Nung (2700 BCE), the "father" of Chinese medicine. He is claimed to have taught Chinese people to practice agriculture, cultivating not only cereals but also

This suggests that today's cannabis differs considerably from the cannabis that was available years ago, both in its effects on mental health and cognitive functions cannabis. Shen Nung is said to have been determined to find alternatives to magic in fighting disease by testing the curative powers of China's plants. The first known Chinese pharmacopoeia, the *Shen Nung Pen Ts'ao Ching* written in the first century BCE, reports all the traditional remedies that were orally handed down for over two thousand years since the legendary Emperor Shen Nung's reign. A preparation of female cannabis flowers was prescribed for all conditions associated with pain, constipation, malaria, and gynecological disorders. It was considered a highly effective herb that was not dangerous. In this ancient text, there is limited reference to psychoactive properties, except that too much cannabis could cause the person to "see demons" (hallucinations) or to "communicate with the spirits." It is likely that the psychoactive use of cannabis was limited to shamans at the time, who because of restrictions of the Chinese Empire, started to leave China heading for India.[3]

In India, cannabis use spread rapidly as a source of happiness and was commonly used in religious rituals as is reported in the sacred text *Artharva Veda*, an ancient collection of holy writings in Sanskrit (around 2000 BCE). The sacred *bhanga* was considered the optimal treatment for anxiety and was used to treat pain, produce anesthesia, reduce spasms and convulsions, and induce hunger.[4] Around 800 BCE in Assyria and in ancient Persia as well as in medieval Arab societies, cannabis was used for its intoxicating and therapeutic effects. Hemp was also grown for its durable fibers. Cannabis use today follows these long traditions of recreational, industrial, and medical use.[5]

The first application of the scientific method to study the pharmacological and toxicological properties of cannabis was by the Irish physician William Brooke O'Shaughnessy, who worked in India and brought the information about cannabis back to the European medical community in the nineteenth century. He conducted both preclinical animal studies of safety and efficacy, as well as human clinical studies. O'Shaughnessy concluded that cannabis was a useful analgesic muscle relaxer and was the most useful treatment known for convulsions.[6]

The French psychiatrist Jacques-Joseph Moreau introduced cannabis to Europeans as a psychoactive drug based on observations made during travel in the Middle East. He adopted the scientific method to detail the psychoactive effects of cannabis. Soon psychotropic use extended beyond therapeutic use, and numerous artists wanted to try cannabis. Moreau provided cannabis to members of the "Club des Hasischins" at the Hotel de Pimodan in Paris (1844–1849), a group of illustrious writers and poets, including Victor Hugo, Alexandre Dumas and Charles Baudelaire. During monthly meetings, Moreau dispensed *dawamesk* (a mixture of hashish, cinnamon, cloves, nutmeg, pistachio, sugar, orange juice, butter, and cantharides) to eminent people who had assembled to ingest the drug. "There are two modes of existence—two modes of life—given to man,"

Moreau mused. "The first one results from our communication with the external world, with the universe. The second one is but the reflection of the self and is fed from its own distinct internal sources. The dream is an in-between land where the external life ends, and the internal life begins." With the aid of hashish, he felt that anyone could enter this in-between land at will.[7] As Moreau studied hashish, he noted a relationship between the amount of the drug taken and its effects. A small dose produced a sense of euphoria and calmness. As the dose increased, attention wandered, ideas appeared at random, minutes seemed like hours, thoughts rushed together, and sensory acuity increased. As the dose increased further, dreams began to flood the brain, like hallucinations of insanity.[8] Indeed, it is now understood that low doses of cannabinoids often produce opposite effects of high doses—a phenomenon described as the biphasic effects of cannabinoids.

Clearly, most effects of cannabis that are only now being studied have been known throughout our human history. For instance, for centuries it has been known that cannabis is effective in treating convulsions (seizures). We now know that CBD is the constituent of cannabis responsible for this antiseizure effect. CBD was approved for treatment of childhood epilepsy by the Federal Drug Administration (FDA) in the United States in 2020, yet it was shown in early clinical trials to be effective in treating epilepsy in 1978.[9] Why has it taken so long to find out how this drug is producing its effects? Prohibition appears to play a role.

Cannabis Prohibition

The fall of medicinal cannabis research in the United States came in 1937 with the enactment of the Marijuana Tax Act, due to the efforts of Harry Anslinger, the supervisor of the Federal Bureau of Narcotics. Cannabis was not banned but had become so prohibitive because of its costs and penalties tied to law violations that all research regarding the potential medical benefits of cannabis was interrupted despite appeals from the American Medical Association. In 1941, cannabis was removed in the United States from the National Formulary and from the Pharmacopeia. In 1961 an international treaty called the Single Convention on Narcotic Drugs placed psychoactive substances into four schedules. Schedule I, the most restrictive, contained drugs viewed to be particularly dangerous for abuse with little therapeutic value. At a subsequent 1971 UN Convention on Psychotropic Substances, the cannabis plant, its resin, extracts and tinctures were all placed in Schedule I, which prohibited all use except for scientific purposes and very limited medical purposes by duly authorized persons. Phytocannabinoids other than THC (such as CBD) were excluded from this control by many countries (e.g., Britain), but the United States and Canada chose to restrict any constituent of cannabis under the same restrictive schedule as THC.[10] This restricted access to cannabis and its constituents has had a negative impact on controlled scientific investigations of cannabis as a drug of abuse and as a potential therapeutic, even considering the shift in public policy with legalization in several states in the United States. Margaret Haney described some of the barriers to research and provided recommendations to bridge the gap between societal use of cannabis and its empirical study.[11] Well-powered, placebo-controlled investigations are critically needed to disentangle pharmacologic efficacy from expectation; yet these studies are nearly impossible to conduct, partially because of the US Drug Enforcement Agency's (DEA's) Schedule I labeling of the cannabis plant and its constituents.[12] It is ironic that as consumers have increasing access to cannabis, US scientists face more regulatory scrutiny and have a limited variety of cannabis and CBD to study. "An important step to addressing many of the barriers facing cannabinoid researchers is to give scientists a schedule I exemption, which would increase the number of randomized control trials and thereby begin to breach the divide between the use of these products and the empirical evidence. Quality science that can weigh societal risks and benefits with data and not hype could then be used to guide nationwide public policy decisions regarding cannabis."[13]

DISCOVERY OF THC AND THE ENDOCANNABINOID SYSTEM

The story of the discovery of THC and the biological system upon which it exerts its effects, the endocannabinoid system, is one of international collaborative scientific adventure at its best. The research of scientists from around the globe revealed a new neurochemical system of the brain and body near the end of the twentieth century.

Discovery of THC

For over a century, researchers attempted to discover and elucidate the structure of the chemical responsible for the characteristic "marijuana high" but mostly were unsuccessful. The first success was in the United Kingdom with the distillation of a red oil extract from Indian charas (a concentrate made from the resin of a live cannabis plant) that produced marijuana-like effects.[1] From the oil a crystalline compound was isolated that was named *cannabinol*. In the 1930s Robert Sidney Cahn and Alexander Todd, also in the United Kingdom, and Roger Adams in the United States, elucidated the structure of cannabinol. Although cannabinol was the first natural cannabinoid extracted in pure form from the cannabis plant, subsequent research revealed that it was not the primary psychoactive chemical in cannabis.

The search continued for the compound in cannabis producing its characteristic psychoactive effects. As we now know, cannabis has more than 125 cannabinoids that have closely related structures and physical properties, so it was difficult to determine which of these several chemicals produced the characteristic "marijuana high." In the 1960s new chromatography separation techniques allowed chemists to isolate and identify chemicals more accurately. Using these modern techniques, Yehiel Ganoi and Raphael Mechoulam finally isolated and identified the active constituent, Δ^9-tetrahydrocannabinol (THC), which produced the "marijuana high."[2] Once discovered, THC was synthesized and made widely available for scientific research. Several thousand papers have since been written describing the effects of THC.

Discovery of Cannabinoid Receptors

Following its discovery, the effects of THC were scientifically evaluated in animal and human studies around the world; yet it was not known how THC produced these effects—what was the mechanism of action producing the "marijuana high"? Because of its lipid chemistry, THC easily diffuses through membranes in the body, so it was first assumed that it acted through a nonspecific membrane associated mechanism. However, as Mechoulam's laboratory synthesized more and more potent analogues of THC, they found that these analogues had very high stereospecificity (i.e., the effect was dependent upon the specific spatial arrangements of the atoms in the molecule). For instance, the synthetic cannabinoid, HU-210, was several thousand times as potent in producing the "marijuana high" as its synthetic mirror image, HU-211. Two very similar synthetic cannabinoids differed greatly in potency only because of the spatial orientation of their atoms. This feature of cannabinoids suggested to researchers that the mechanism of action must be one of binding to a specific receptor (like a key fitting into a lock of a specific shape).

In 1988, Allyn Howlett's research group in St. Louis, discovered the specific receptor for THC, which they called a *cannabinoid-1 (CB$_1$)-receptor*.[3] Later a second, cannabinoid receptor (CB$_2$), mostly found in the peripheral nervous system (outside the brain and spinal cord—central nervous system), was identified in the spleen.[4] The CB$_1$ receptor was initially believed to be expressed only in the central nervous system (CNS) and was considered a brain cannabinoid receptor; however, it is now clear that it is

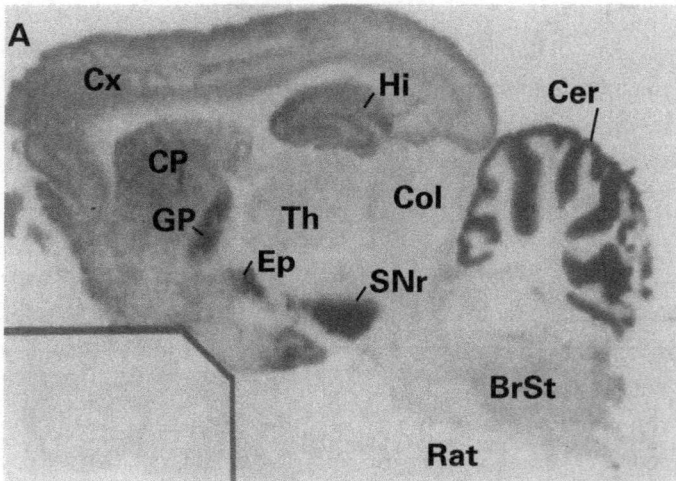

Figure 1 Distribution of brain CB₁ receptors shown by autoradiography of CP 55,940 (potent CB₁ agonist) binding in the rat brain. Gray levels represent relative levels of receptor densities. Sagittal section of rat brain. BrSt: brainstem; Cer: cerebellum; Col: colliculi; CP: caudate-putamen; Cx: cerebral cortex; Ep: entopeduncular nucleus; GP: globus pallidus; H: hippocampus; SNr: substantia nigra; Th: thalamus. Based on Herkenham et al. 1990.

also present in several peripheral organs (but at a lower expression level). In the brain, we now understand that CB₁ receptors are expressed in a wide variety of regions; CB₁ receptors are among the most widely expressed receptors in the brain. Figure 1 presents the distribution of brain CB₁ receptors shown by autoradiography of a CB₁ agonist binding in a rat brain.[5] The highest densities of CB₁ receptors in the rat brain are found in the basal ganglia, substantia nigra, globus pallidus, hippocampus, and the cerebellum, but not in the brainstem. Because the brainstem regulates autonomic functions necessary for survival (cardiovascular, respiration), the paucity of CB_1 receptors in it explains the low toxicity of THC. As is not the case with morphine, no one has ever died due to respiratory distress of an overdose of cannabis.

Discovery of Endocannabinoids

But why does our brain contain receptors for THC, a chemical found in a plant? Earlier in the 1970s a similar question was asked when the opiate receptor was discovered in the rat brain. The discovery of a receptor for morphine led to a search for endogenous chemicals that act like morphine at this receptor. This search led to the discovery of a family of peptides known as endorphins (endogenous morphines).[6] Similarly, the discovery of cannabinoid receptors prompted a search for naturally occurring cannabinoids that acted on the CB_1 and CB_2 receptors.

The plant constituent, THC, which just happens to bind to these receptors is a lipid compound; hence it was assumed that any possible endogenous cannabinoid molecules (endocannabinoids) would also be lipids. Mechoulam and his colleagues isolated and identified two such

The discovery of cannabinoid receptors prompted a search for naturally occurring cannabinoids that acted on the CB_1 and CB_2 receptors.

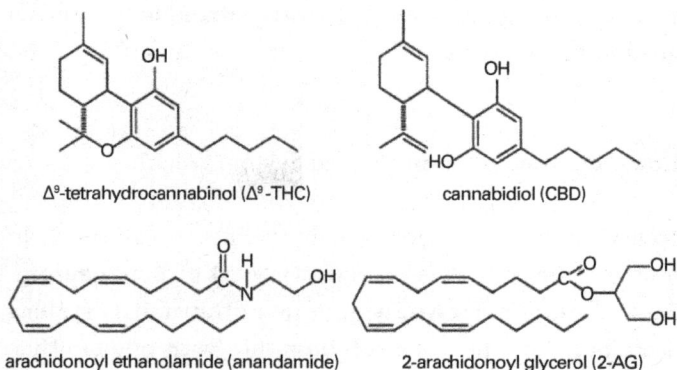

Figure 2 Structures of the plant cannabinoids Δ⁹-tetrahydrocannabinol and cannabidiol and the endogenous cannabinoids anandamide and 2-arachidonoyl glycerol.

compounds—one from brain tissue (anandamide, abbreviated AEA, its name derived from the Sanskrit word *ananda*, meaning, "supreme joy")[7] and one from peripheral tissues (2-arachidonoyl glycerol (abbreviated 2-AG).[8] The latter, 2-AG, was also isolated from brain tissue by Takayuki Sugiura and others and is therefore found in both the brain and peripheral tissues.[9] There have since been thousands of articles published about these two compounds and their functions. Their discovery has revealed a new, unique biological system called the *endocannabinoid system*. Figure 2 presents the structures of the plant cannabinoids, Δ⁹-tetrahydrocannabinol and cannabidiol, and the endogenous cannabinoids, anandamide and 2-arachidonoyl glycerol.

How Does the Endocannabinoid System Work?

To explain how the endocannabinoid system works, it is first necessary to give a brief overview of how neurons work. Figure 3 portrays a generic neurotransmitter system describing how neurons communicate with one another using neurotransmitters in the brain and body. We see that the sending neuron (presynaptic neuron) and the receiving neuron (postsynaptic neuron) make chemical contact

Generic Neurotransmitter System

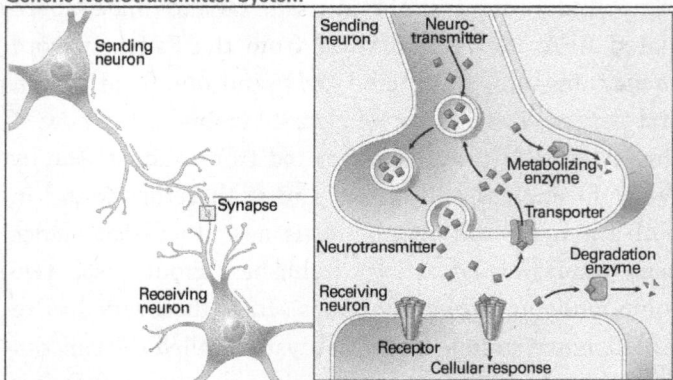

Figure 3   A generic neurotransmitter system.

at a site called a synapse (a gap between neurons filled with extracellular fluid). The synapse is enlarged in the right half of the figure. The sending neuron transmits an electrical signal (an action potential) to its terminal ending at the presynapse where neurotransmitters are located and stored in vesicles waiting to be released into the synapse when the electrical impulse arrives. When released from the presynapse, the neurotransmitter diffuses in an anterograde (forward) manner across the synapse (through extracellular fluid), ultimately binding with postsynaptic receptors located on the membrane of the receiving (postsynaptic) neuron, thereby triggering a cellular response on the receiving neuron. This generic process of synaptic transmission occurs for neurons containing the classical neurotransmitters, such as glutamate, GABA, dopamine, norepinephrine, serotonin, and acetylcholine. Once the postsynaptic receptors on the receiving neuron have been activated, the action of the neurotransmitter is terminated. The termination of action is accomplished by the release of the neurotransmitter from the receptor back into the synapse. The chemical diffuses back across the synapse and is transported back into the presynaptic neuron where it is ultimately repackaged into the vesicles awaiting to be released when the neuron fires again. Almost miraculously, this entire process occurs in less than a millisecond!

Endocannabinoids (AEA and 2-AG) act differently at the synapse than classical neurotransmitters—they are

Endocannabinoids (AEA and 2-AG) act differently at the synapse than classical neurotransmitters—they are produced and released by the postsynaptic (receiving) neuron, not the presynaptic (sending) neuron.

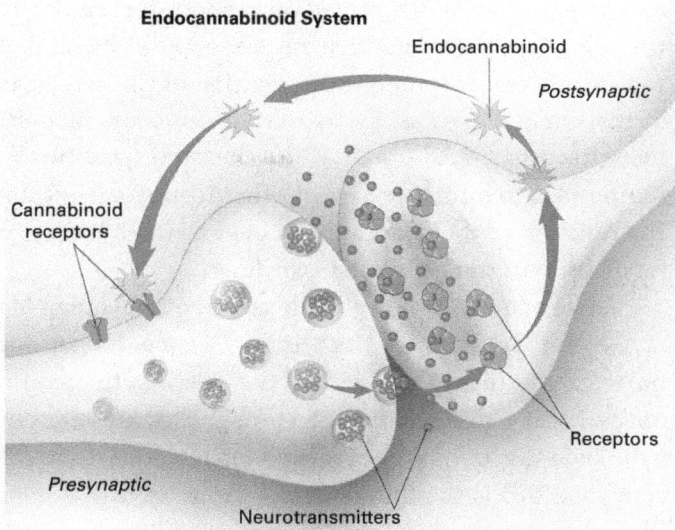

Figure 4 The endocannabinoid system.

produced and released by the postsynaptic (receiving) neuron, not the presynaptic (sending) neuron. As seen in figure 4, in contrast with most neurotransmitters, AEA and 2-AG are retrograde (backward) messengers released from the postsynaptic membrane. They are not stored in vesicles in the presynaptic neuron but are instead synthesized by the postsynaptic neuron when and where they are needed. They are then released by the postsynaptic neuron into the synapse and travel back across the synapse to act on CB_1 receptors located on the terminal ending of the presynaptic neurons that release other classical neurotransmitters. This unique distribution of CB_1 receptors at the synapse gives a clue as to their function. The ultimate function of AEA or 2-AG on these presynaptic CB_1 receptors is to inhibit release of other neurotransmitters. The function of the endocannabinoid system is simply to regulate neurotransmission in the brain.

As seen in figure 4, the synthesis of both anandamide and 2-AG is triggered by the action of the neurotransmitter (e.g., glutamate) released by the presynaptic neuron on excitatory postsynaptic receptors. That is, endocannabinoids are only produced in brain regions that are activated; this is the "on-demand" nature of this system. The action of an excitatory neurotransmitter (such as glutamate) at the receptor results in an influx of calcium across the postsynaptic membrane. This influx of calcium induces postsynaptic biosynthesis and release of AEA and 2-AG into the synapse. 2-AG is formed from its precursor, diacylglycerol (DAG), and AEA is formed from its precursor, NAPE (N-arachidonlyl phosphatidylethanolamide). When released into the synapse, 2-AG and AEA diffuse in a retrograde fashion across the synapse to bind with CB_1 receptors on the membranes of the presynaptic neuron. The CB_1 receptor is inhibitory and results in the reduction of the release of the activating neurotransmitter. CB_1 receptors act in the same manner to reduce the release of the primary inhibitory neurotransmitter (GABA) in GABAergic neurons. Therefore, CB_1 activation on glutamatergic neurons reduces excitation, and CB_1 activation on GABAergic neurons reduces inhibition. In this way, the endocannabinoid system acts as a homeostatic regulator of neural activity—it fine-tunes the action of other neurotransmitters!

The CB_1 receptor is also activated by THC in this manner; that is, THC is an agonist of the CB_1 receptor in the brain. However, the action of THC is longer than that of endocannabinoids. Whereas THC is metabolized over several hours and excreted (or stored as one of its metabolites), endocannabinoids are removed from the synapse within minutes by a membrane transport system. AEA diffuses back across the synapse through the postsynaptic membrane where it is then deactivated by the enzyme FAAH (fatty acid amide hydrolase). On the other hand, 2-AG is transported across the presynaptic membrane where it is deactivated by the enzyme MAGL (monoacylglycerol lipase). Suppression of these enzymes prolongs the action of the endocannabinoids at the specific site at which they are produced; drugs with suppress these enzymes are being developed as a new way of manipulating the endocannabinoid system for medical benefits as we shall see in the chapters that follow. Unlike THC, which floods the brain acting on all CB_1 receptors, the natural release of endocannabinoids occurs only in those brain regions activated at the moment, when and where they are needed, and these enzyme inhibitors prolong this specific action.

The endocannabinoid receptors, the endocannabinoids, and their biosynthetic and biodegrading enzymes constitute what is known as the endocannabinoid system. Although 2-AG and AEA broadly act in a similar manner in the brain, the concentration and efficacy of 2-AG in the brain is higher than AEA. Yet AEA appears to be activated when physiological systems are stressed.

CB_2 receptors

CB_2 receptors are primarily present in peripheral cells of the immune system acting in a protective manner. The action of 2-AG and AEA on CB_2 receptors is primarily the mechanism for their immunosuppressive effects. CB_2 receptors have been identified at a much lower level than CB_1 receptors in the CNS, generally under pathological conditions on microglial cells under conditions of stress. CB_2 receptor expression is enhanced in the CNS and other tissues under pathological conditions. Whether CB_2 receptors are present in a noninjured brain is a matter of debate. It has been argued that the CB_2 receptor is a part of a general protective system.[10] Several CB_2 agonists have been synthesized to determine their neuroprotective potential. Because CB_2 receptor agonists do not cause the psychoactive effects associated with CB_1 agonists, CB_2 agonists may be useful drugs to treat pain (see chapter 9),

In this way, the endocannabinoid system acts as a homeostatic regulator of neural activity—it fine-tunes the action of other neurotransmitters!

neurodegenerative diseases (see chapter 10), cardiovascular disease, and liver disease.[11]

Transient Receptor Potential V1 (TRPV1)

AEA also acts as an agonist of an intracellular postsynaptic receptor called a transient receptor potential V1 (TRPV1), which is also activated by capsaicin (the hot stuff in chili peppers); these receptors are often called capsaicin receptors that can suppress pain sensitivity (see chapter 9). The binding site is located on the inner face of the postsynaptic cell membrane suggesting that when AEA is biosynthesized by the cells that express TRPV1, it will activate the receptor before being released into the synapse. THC and synthetic CB_1 and CB_2 agonists also interact with TRPV1, but usually with a lower relative intrinsic activity and potency than is exhibited by AEA.

G-Protein Receptor 55 (GPR55)

The GPR55 receptor was considered a third cannabinoid receptor for some time because some effects of compounds thought to be specific for CB_1 or CB_2 receptors appeared to also act independently of these receptors. We now know that a different bioactive ligand, lysophosphatidylinositol (LPI), acts as an agonist on the GPR55 receptor. CBD acts at multiple targets, including acting as an antagonist on the GPR55 receptor. The CB_1 antagonist/inverse agonist, rimonabant, acts as an agonist at the GPR55 receptor, but it requires a higher concentration than is required to antagonize the CB_1 receptor. GPR55 receptors are located on glutamatergic axon terminals, where they increase glutamate release when the neuron fires. Because CBD effectively blocks GPR55 activation, it has anticonvulsant activity by reducing excess glutamate release from the hyperactive excitatory neurons during epileptic seizures.

GPR55 is also involved in weight gain—circulating LPI levels (the endogenous agonist of GPR55 receptors) are elevated in obese patients and are correlated positively with body weight.

Manipulations of the Endocannabinoid System

CB_1 and CB_2 receptor agonists and antagonists. Cannabinoid receptors (CB_1 and CB_2) contain a primary binding site (orthosteric binding site) to which the ligand (i.e., AEA or 2-AG) binds and activates that receptor. The plant cannabinoid, THC, also binds with these receptors, as well as synthetic agonists for the CB_1 or CB_2 receptor by mimicking the shape of the natural ligand. The synthetic CB_1/CB_2 agonists, nabilone (Cesamet©) and dronabinol (Marinol©) were licensed and approved by the Federal Drug Administration (FDA) in the United States in the 1980s as medications for chemotherapy-induced nausea, vomiting, and anorexia. Because activation of peripheral CB_2 receptors does not produce psychoactivity but does produce analgesia in preclinical animal models of pain, there has been considerable recent interest in the development of CB_2 agonists particularly for the treatment of chronic pain.

CB_1 or CB_2 antagonists also bind to these primary orthosteric binding sites, but instead of activating the site, they block the site from binding by an active agonist. As well, a CB_1 or CB_2 inverse agonist binds with the primary site but produces an effect opposite of that of the agonist—thereby acting in a manner opposite to the agonist. Rimonabant, originally devised as an antiobesity drug (see chapter 7), acts as an inverse agonist/antagonist of the CB_1 receptor. CB_1 and CB_2 receptor agonists have been useful research tools for investigators interested in understanding the functioning of the endocannabinoid system.

Allosteric modulators of the CB_1 and CB_2 receptors. Not only do cannabinoid receptors have a primary or orthosteric binding site to which the natural ligand (AEA and 2-AG) binds, but they also contain an allosteric binding site. A molecule that binds to an allosteric site changes the shape of the main receptor so that the intensity of the effect of an agonist binding to the orthosteric site is modified. However, allosteric modulators that bind to these allosteric sites have no physiological effects on their own. Allosteric modulators can be either negative, resulting in a reduced intensity of the agonist at the primary binding site, or positive, resulting in an increased intensity of the agonist at the primary binding site. Recent evidence indicates that CBD acts as a negative allosteric modulator of CB_1 and CB_2 receptors, thereby reducing the effects of THC (as well as AEA and 2-AG) at the orthosteric binding site.[12]

Inhibitors of endocannabinoid degrading enzymes. Following their endogenous release, endocannabinoids are rapidly enzymatically degraded. AEA is transported back across the postsynaptic membrane where fatty acid binding (FAB) proteins transport it to the intracellular enzyme, Fatty acid amide hydrolase (FAAH), for inactivation. 2-AG is also inactivated enzymatically but in the presynaptic neuron by monoacylglycerol lipase (MAGL). Suppression of these enzymes prolongs the activity of the endocannabinoids.

We shall see that new pharmaceutical compounds are being developed that suppress FAAH or MAGL, thereby increasing (or "boosting") the duration of action (from a few minutes to up to twenty-four hours) and the endogenous level of AEA and 2-AG, respectively. The rationale behind this approach to drug development is based on the mechanism of AEA or 2-AG formation and release, which is known to occur when and where they are needed (via an on-demand system). As noted above these endocannabinoids are not stored in synaptic vesicles (as classical neurotransmitters), but instead are synthesized in the postsynaptic neuron and released into the synapse after neuronal activation. Therefore, inhibition of AEA or

New pharmaceutical compounds are being developed that suppress FAAH or MAGL, thereby increasing (or "boosting") the duration of action (from a few minutes to up to twenty-four hours) and the endogenous level of AEA and 2-AG, respectively.

2-AG deactivation would enhance CB_1 activation where the endocannabinoid levels are highest, but not globally throughout the brain and body (as with the systemic administration of a CB_1 agonist, like THC). Such an approach produces elevated levels of endocannabinoids, specifically in those regions for as long as twenty-four hours. Treatments that boost the natural endocannabinoid system do not produce the same psychoactive and other side effects as global CB_1 agonists. FAAH and MAGL inhibitors are being developed and tested for several conditions including pain, inflammation, cannabis withdrawal disorder, anxiety, depression, posttraumatic stress disorder, appetite enhancement, addiction, neurodegenerative disorders, and nausea.

CANNABIS CONSTITUENTS
AND PHARMACOLOGY

The term *cannabinoid* usually refers to the chemical substances isolated from the cannabis plant that possess the typical C_{21} terpenophenolic skeleton. These molecules are called phytocannabinoids because they come from the plant; well over a hundred of these have been identified.

Phytocannabinoids show different affinities for the CB_1 and CB_2 receptors, with other molecular targets also being identified.[1] Investigation of the therapeutic potential of these cannabis constituents has been largely limited to THC and CBD in the past, although there has been a recent interest in the study of other phytocannabinoids. Synergistic, or at times antagonistic, effects of these cannabis constituents occur because of their different actions at different receptors of the body. This is called the "entourage effect."[2]

Synergistic, or at times antagonistic, effects of these cannabis constituents occur because of their different actions at different receptors of the body. This is called the "entourage effect".

Cannabis Constituents

THC-type cannabinoids. Δ^9-THC (which we call THC throughout the book) and cannabinoids similar in structure to THC have been the most extensively studied cannabinoids. THC is present in most parts of the plant but is most highly concentrated in the droplets of sticky resin produced by glands at the base of the female flowers. After its identification by Yehiel Ganoi and Raphael Mechoulam, THC was tested for activity in rhesus monkeys, dogs, gerbils, mice and rats, and was found to produce the typical psychoactive effects of cannabis.[3] Some of the effects produced by THC in these early animal studies were severe motor disturbances, redness of the mucous membrane that covers the eyeball, slow movements, decline of aggression, sleepy state, and decreased spontaneous locomotion. The mouse tetrad assay was then developed to identify compounds with THC-like activity. The behaviors of the tetrad include catalepsy (muscular rigidity), reduced motor activity, reduced body temperature and pain relief.[4] The mouse tetrad assay is commonly used to screen for psychotropic cannabinoids. THC is a partial agonist at the CB_1 and CB_2 receptors. Small quantities of another THC compound, Δ^8-THC, have also been identified in cannabis, which like Δ^9-THC is psychoactive and acts as a partial agonist at the CB_1 and CB_2 receptors. Δ^8-THC has been used in children undergoing chemotherapy to prevent vomiting, with few reported side effects.[5] Because it is not the primary THC compound in cannabis, it will generally not be discussed in this book—any reference to THC will represent Δ^9-THC. The raw cannabis plant does not contain THC, but instead contains its carboxylic acid precursor THCA (both Δ^9-THC acid and Δ^8-THC acid). THCA is decarboxylated to THC by heating (smoking and baking). THCA is not psychoactive because it has low affinity for the CB_1 receptor. Tetrahydrocannabivarin (THCV) is also found in the cannabis plant and acts as an antagonist, rather than an agonist at the CB_1 receptor. As a CB_1 antagonist, THCV reduces food intake and body weight (like rimonabant see chapter 7).

Cannabinol. Cannabinol (CBN) was the first cannabinoid to be isolated from cannabis extract in the late nineteenth century, and its structure was identified in the 1930s by Cahn.[6] Further research however revealed that it does not produce the typical THC effects. It is a weak agonist of CB_1 and CB_2 receptors, however, there has been little recent research on its effects.

Cannabidiol-type cannabinoids. The primary nonintoxicating cannabinoid of cannabis (particularly hemp) is CBD, first isolated by Adams and colleagues.[7] Mechoulam subsequently isolated CBD from Lebanese hashish and established its structure.[8] CBD lacks the psychotropic effects of THC and has considerable therapeutic potential.[9] Unlike THC, CBD does not activate CB_1 or CB_2 receptors,

The raw cannabis plant does not contain THC, but instead contains its carboxylic acid precursor THCA (both Δ^9-THC acid and Δ^8-THC acid). THCA is decarboxylated to THC by heating (smoking and baking).

explaining its lack of psychoactivity. Instead, CBD acts through multiple mechanisms. CBD acts as an agonist at a specific serotonin receptor, 5-HT_{1A}, and TRPV1 channels (see chapter 2). CBD also acts as an antagonist at the GPR55 and as a negative allosteric modulator of both the CB_1 and CB_2 receptors (see chapter 2[10]) and may negatively affect signaling at adenosine A_{2A} receptors upon which caffeine also acts. An oral solution based on pure plant-derived CBD (Epidiolex®) has been recently approved in the United States for the treatment of the childhood epileptic seizures (Dravet syndrome and Lennox-Gastaut syndrome) in patients two years of age and older (see chapter 10). CBD has several other potential beneficial effects that will be discussed throughout the book.

Although the concentration of THC has dramatically increased in the past decade, over the past few years more high-CBD products have been found among illicit cannabis products seized by the Drug Enforcement Agency in the United States.[11] Of the confiscated samples analyzed, the mean THC concentration increased from 10 percent in 2009 to 14 percent in 2019, a with a mean THC:CBD ratio increasing from 25 in 2009 to 105 in 2017. Likely because of evolving interest in the potential benefits of CBD, the THC:CBD ratio decreased back to 25 in 2019, suggesting a trend in the production of high-CBD products in the illegal market.

The Farm Bill (2018) in the United States removed hemp from the list of controlled substances in the Controlled Substances Act, meaning that if CBD is derived from hemp (cannabis with less than 0.3 percent THC), it does not fall under the jurisdiction of the DEA. This means, however, that CBD derived from cannabis with a higher THC content remains as a Schedule I drug (however, as of this writing there are ongoing hearings by the DEA to consider rescheduling cannabis to Schedule III). Clinical studies using CBD must still be approved by the FDA, making RCTs difficult for researchers to conduct. It is rather ironic that as consumers have greater access to CBD, researchers continue to face greater scrutiny in conducting necessary RCTs. Curiously, over-the-counter CBD products have been evaluated in research laboratories, and it has been found that 69 percent of them are incorrectly labeled as "CBD-only" with 49 percent underlabeled and 21 percent also contained THC.[12]

Like THC, CBD is not present in the raw cannabis plant. Cannabidiolic acid (CBDA) is the precursor of CBD that is present in the fresh cannabis plant (particularly in its industrial hemp forms). It slowly decarboxylates (i.e., loses its acidic function) in response to heating (e.g., when cannabis is smoked). CBDA has recently received considerable investigation for its medicinal effects, which is discussed in later chapters. A related constituent, Cannabidivarin

Over the past few years, more high-CBD products have been found among illicit cannabis products seized by the Drug Enforcement Agency in the United States.

(CBDV), also lacks psychoactive properties, but has potential efficacy for controlling seizures.

Cannabigerol. Cannabigerol (CBG) is often referred to as the mother of all cannabinoids, as it is the decarboxylated form of CBG acid (CBGA) that is the parent molecule from which other cannabinoids are synthesized, including THCA and CBDA. CBG is normally a minor constituent of cannabis that is typically consumed. During plant growth, most of the CBG is converted to other cannabinoids, leaving about 1 percent of CBG in the plant (however, some strains produce larger amounts of CBG). Preclinical research indicates that CBG reduces anxiety and depression in rodents. There has been a recent double-blind placebo-controlled crossover field trial with CBG in healthy adults demonstrating the potential of CBG (oral 20 mg hemp–derived CBG or placebo tincture) to reduce anxiety and stress and elevate mood.[13] CBG may represent a novel option to reduce anxiety and stress in healthy adults.

Terpenes. Terpenes in cannabis produce the plant's aroma and reported "flavor." Terpenes are the odorous compounds present in essential oils. More than a hundred terpenes have been identified in cannabis. Preclinical evidence indicates that terpenes may have therapeutic potential. D-limonene, β-myrcene, and α-pinene are some of the most common terpenes in cannabis. The literature suggests that terpenes may also act synergistically with cannabinoids to produce beneficial effects.[14] Indeed, combinations of cannabinoids and terpenes could provide promising therapeutic tools, which may ultimately reveal why people attribute relief from certain symptoms to specific cannabis strains.

Cannabis Pharmacology

Routes of administration. Cannabis is self-administered by smoking, by vaporization, by eating, by sublingual (under the tongue) and oral application of extracts, and by topical application of a salve. Vaporization (heating until volitive active cannabinoids are vaporized) may reduce combustion products in the inhalant, making it a safer method than smoking, although some of the toxins or carcinogens may still be present in the vapor. Smoking or vaporization produces more rapid onset (within minutes) of the intoxicating effects of cannabis. This allows the user to regulate the desired effect by taking many or fewer puffs (called titration). When cannabis is taken orally, the onset of the effects is slower (up to 90 minutes) and run the risk of unpredictable effects because the dose is more difficult to control. Care must be taken to keep edibles out of the reach of children who may mistake them for candy. Cannabinoids administered topically must be prepared in special lipid vehicles to allow absorption through the skin—however, the bioavailability of topically applied cannabinoids has been little studied.

Bioavailability and metabolism of THC. Most people using smoked or orally ingested cannabis for medical purposes report using between 10 and 20 gm of cannabis per week, with a median daily dose of 2 gm of smoked cannabis, 1.5 g of vaporized cannabis, or 1.5 gm of oral cannabis in food or teas.[15] A typical marijuana cigarette contains between 0.5 and 1 gm of cannabis plant material, which may vary in THC content between 7.5 and 22.5 mg (between 7 percent and 30 percent) and in CBD content between 0 and 180 mg (0–24 percent). The minimum dose of THC necessary to produce pharmacological effect in humans is between 2 and 22 mg.[16]

The term *bioavailability* refers to the fraction of an administered dose of an unchanged drug that reaches systemic circulation. When a drug is administered directly into the bloodstream intravenously, its bioavailability is 100 percent. When a drug is administered by inhalation, orally, or transdermally (by application to the skin), its bioavailability is decreased because of incomplete absorption into the blood or metabolism by liver enzymes. When cannabis is smoked, approximately 25 percent is bioavailable with a peak plasma concentration (C_{max}) in 6–10 min. Because the effects are rapid, smoking allows the user to titrate (adjust the dose for maximum effect) the dose of

THC. Oral ingestion of cannabis results in approximately 6 percent bioavailability because it must pass through the gut and the liver where enzymes act to metabolize it ("first-pass metabolism") before entering the systemic blood stream; however, when it is suspended in an oil (such as sesame oil) bioavailability improves. When orally consumed, time to peak plasma (C_{max}) is about 2–6 hours.

THC is metabolized in the liver by hepatic cytochrome P450 and several other drug-metabolizing enzymes to an active equipotent metabolite, 11-hydroxy-THC, and an inactive metabolite during cannabis smoking.[17] After smoking cannabis, the active metabolite of THC is at about 10 percent of the level of THC in the blood, and peaks about 15 minutes after the start of smoking. After oral administration of cannabis, the active metabolite is about 3 times as high as after smoking and peaks at 2–4 hours. The greater concentration of the active metabolite following oral consumption is the result of first-pass metabolism—that is, when orally consumed the THC enters the liver where it is metabolized prior to being absorbed into the blood stream, unlike inhaled THC.

The concentration of THC in the blood decreases rapidly after smoking because of tissue distribution, liver metabolism, and urinary and fecal excretion. Because THC is highly lipophilic, it is rapidly taken up by tissues with high blood flow, including the heart, lungs, brain, and liver. Tissues with less blood flow accumulate THC more slowly and release it over a longer period. THC stored in fat in chronic frequent cannabis smokers can be released into the blood for days.[18] Within 5 days, 80–90 percent of a THC dose is excreted. Detection times in urine after smoking a 3.5 percent THC cigarette range from 2 to 5 days for occasional cannabis smokers but can extend to weeks in chronic daily cannabis smokers.[19] The estimates of terminal half-life (time to excrete half of a dose) of THC in humans have increased progressively as analytic methods have become more sensitive. With repeated use, people become tolerant to THC because chronic activation of the CB_1 receptor produces receptor desensitization or downregulation.

Bioavailability and metabolism of CBD. Only a few published studies report bioavailability of CBD in humans. CBD is generally administered orally either as a capsule or dissolved in an oil solution in clinical trials or research studies with humans. CBD is a highly lipophilic drug and is poorly absorbed in the aqueous environment of the gut. As well, when administered orally, it undergoes first-pass metabolism in the liver before it enters the circulatory system, and a large proportion (33 percent) is excreted unchanged in the feces. It has been estimated that the bioavailability of 20 mg of CBD is only 6 percent when given orally in a fasted state; however, bioavailability increases up to four times when administered in a fed state.[20] A current focus of pharmacological research with CBD is to find new vehicles (solutions in which CBD is mixed) that increase bioavailability, such as long chain fatty acids high in oleic acid or nano-lipospheres, with both showing higher bioavailability compared with generic CBD.[21] Once in the blood stream, CBD is rapidly distributed to bodily tissue, including the brain, but like THC it preferentially accumulates in adipose tissue because of its lipophilicity.

When administered orally, CBD is extensively metabolized in the liver. Seven human cytochrome P450 enzymes are capable of metabolizing CBD, which are also involved in metabolizing other clinically important drugs. This suggests that CBD may modify the metabolism of other drugs that a patient may be taking. In fact, the metabolism of THC may be inhibited by CBD.[22] In clinical trials for epilepsy, CBD is given as an adjunct therapy in addition to the patient's normal antiepileptic drugs. This coadministration of CBD resulted in an increased plasma concentration of clobazam's (a typical antiepileptic drug) major active metabolite, likely the result of inhibition of clobazam metabolism by CBD.[23] However, subsequent clinical trials verified that such drug-drug interactions cannot account for the antiseizure effects of CBD.[24] A recent comprehensive review describes the interactions between CBD and other medications, illicit substances, and alcohol.[25]

CBD safety. Several reviews have evaluated the safety and toxicity of CBD.[26] Generally, preclinical studies show CBD to be relatively nontoxic and safe. However, less is understood about the long-term use of higher doses in humans used to treat epilepsy (see chapter 11) and schizophrenia (see chapter 4). As noted above, there is a potential for CBD to be associated with drug interactions through inhibition of some cytochrome P450 liver enzymes. The impact of CBD alone and in combination with other common medications requires continued monitoring. Most of the commonly reported side effects in human clinical trials with CBD include fatigue, diarrhea, and changes of weight or appetite. A recent meta-analysis of Randomized Control Trials (RCTs) with CBD suggests the likelihood of voluntary withdrawal from a study because of side effects is greater with high-dose CBD (1,400 to 3,000 mg/day) than placebo.[27] The doses (5–20 mg/day) of CBD provided by health and food supplements are generally much lower, so incidence of adverse side effects is likely to be much lower, notwithstanding the issue of poor labeling of over-the-counter products.[28]

CBD/THC interactions. The two constituents of cannabis that have gained the most research attention are THC and CBD. Although CBD is nonintoxicating, it may interact with the effects of THC. This interaction is complex such that under some conditions CBD reduces the effects of THC, but under other conditions CBD potentiates the effects of THC.

CBD has been shown to reduce some harmful effects of high doses of THC.[29] High doses of THC can produce acute psychosis and anxiety (see chapter 5), as well as cognitive impairment (see chapter 6). When coadministered, CBD has been reported to reverse these harmful effects of high-potency THC in both human and animal studies. A mechanism for this reversal has been suggested to be the opposing actions of THC and CBD at the CB_1 receptor in the brain. As a negative allosteric modulator of the CB_1 receptor, CBD has the effect of reducing the binding of CB_1 agonists such as THC.[30]

A byproduct of the selective breeding for high potency strains of cannabis has been the breeding out of CBD such that very low levels are typical in street cannabis.[31] It has thus been suggested that the absence of CBD in cannabis of high THC potency may contribute to the psychosis-like outcomes. A harm minimization strategy has recommended reinstatement of CBD into cannabis plant material and products. In high-potency cannabis, CBD concentrations have declined substantially.

The amelioration of adverse effects of THC by CBD may be dose-dependent. Indeed, a recent randomized clinical trial reported that when administered by inhalation (vaporization), high doses of CBD reduced, but low doses of CBD enhanced, the intoxicating effects of THC.[32] As well, preclinical evidence suggests that CBD may potentiate the pain-relieving effect of low doses of THC and the nausea/vomiting-relieving effect of low doses of THC, suggesting that the combined use of low doses of these plant cannabinoids may increase treatment efficacy.[33] The potentiating effects of THC by CBD appear to be mediated by non-CB_1R mechanisms.

Most of previous RCTs have used inhaled cannabis, and the results may not translate to edibles that undergo first-pass metabolism by drug-metabolizing P450 enzymes in the intestine and liver before reaching the systemic circulation. Thereby, CBD may compete with the metabolism of THC by enzyme inhibition in the liver. Oral administration of CBD with THC has thus been shown to raise THC concentrations in blood and brain, prolonging THC activity in the CNS.[34] A very recent randomized control trial in humans found that oral administration of a high dose of CBD with a moderate dose of THC enhanced the anxiety-provoking and cognitive-impairing effect of THC, presumably by inhibiting the metabolism of THC and its psychoactive metabolite, 11-OH-THC.[35] As well, very high oral CBD doses of 300–600 mg/day in humans used to treat childhood epilepsy and schizophrenia have also been shown to inhibit the metabolism of other drugs acted upon by P450 liver enzymes—thereby elevating the levels of accompanying drugs consumed. These doses of CBD are much higher than those available in over-the-counter products. A better understanding of cannabinoid interactions with other drugs is necessary to inform clinical and policy decision-making regarding the therapeutic and nontherapeutic use of cannabinoids.

Interactions between THC and CBD appear to be very complex with a wide range of potential mechanisms. What seems to be important is that these differential effects may depend upon absolute dose, ratio of CBD:THC, route of administration, and whether CBD is given prior to or simultaneous with THC, but no definite conclusion has yet emerged.

Nabiximols. Nabiximols for clinical use is a combination of CBD and THC generally in a 1:1 concentration of extracts from the dried flowers of two select cannabis strains: Tetranabinex yields high THC content, and the Nabidiolex yields a high-CBD content. Nabiximols is composed mainly of THC and CBD (70 percent) but also contains other phytocannabinoids derived from the plant material.[36] Sativex[©] (GW Pharmaceuticals) is nabixomols administered by oral spray (2.5 mg THC: 2.7 mg CBD) sublingually (under the tongue). Sativex is approved for clinical use in multiple sclerosis (MS, for spasticity and pain) and cancer pain in Britian, throughout Europe, in Canada, and in many other countries, but has not been approved in the United States by the FDA for these uses. The 1:1 combination of THC: CBD was selected to allow higher doses of THC without increasing the risk of adverse side effects, as CBD may act to antagonize some of the psychoactive and sedative effects of THC without interfering with intended effects of THC, such as muscle relaxation and reduction of spasticity in MS. Nabiximols does not produce significant adverse cognitive or psychomotor side effects and show a lower abuse potential than dronabinol (synthetic THC) at lower doses.[37] However, both medications at the highest doses tested exhibited some abuse potential, defined as self-reported liking for a drug compared with placebo. Therefore, careful monitoring of abuse and of aberrant medication-related behaviors during clinical treatment with nabiximols is warranted.

Potential Risks Associated with Recreational Cannabis

The potential risks associated with recreational cannabis are most likely associated with the intoxicating constituent, THC. This is particularly pertinent concerning the proliferation of strains and extracted tinctures of high-potency THC, which are increasingly becoming available in market. The most common concerns about potential risks of cannabis are discussed in chapters that follow, and these include: the potential for addiction (chapter 4), the potential of cannabis to precipitate psychosis (chapter 5), and the potential of cannabis to impair brain development and learning (chapter 6). Several additional risks have been identified:

1) Respiratory risks: Many of the potential carcinogens found in tobacco smoke are found in cannabis smoke.

Regular cannabis use by inhalation is associated with greater risk of respiratory disorders.[38] These respiratory risks may be reduced by using vaporizers.[39]

2) Cardiovascular risks: Cannabis can increase heart rate and supine blood pressure but decrease blood pressure upon standing. In rare cases, it has been reported to trigger acute myocardial infarction.[40]

3) Driving: As there are no simple and accurate roadside tests for cannabis intoxication, driving while intoxicated is another risk factor for cannabis. Driving while intoxicated with cannabis appears to increase the risk of being in a motor vehicle accident. Experimental data using driving simulators show that people attempt to compensate by driving more slowly after smoking cannabis, but that control deteriorates with increasing task complexity. Cannabis smoking impairs cognitive function, increases reaction times, and increases lane swerving in driving simulators.[41]

CANNABINOIDS: REWARD AND ADDICTION

How does cannabis produce euphoria? Can cannabis be considered an abused drug? Is it addictive? How does it and the endocannabinoid system modify the rewarding properties of other drugs? These are all questions that researchers are trying to answer.

Clearly heroin, nicotine, amphetamine, and alcohol have the potential to produce addiction in humans. These drugs have all been shown to engage the mesolimbic dopamine system of the brain. THC and endocannabinoids also activate this system.[1] The term *addiction* generally refers to pathological drug-taking ranging from mild to severe. At first, intake of a drug gradually escalates from occasional use to more sustained and regular intake; later discontinuing the drug becomes difficult and the individual experiences "drug withdrawal effects." This latter phase is characterized by loss of control of drug use. More drug is taken, more time is devoted to drug-seeking and -taking activities, and the person cannot stop using the drug even though experiencing adverse consequences. According to the American Psychiatric Association's *Diagnostic and Statistical Manual of Mental Disorders* (*DSM*), "substance-use disorders" replaces the term "addiction," which is commonly used only to refer to the severe final state of a substance use disorder. Thus, there is a continuum in the severity of disorders related to drug use.

Cannabis Use Disorder

It was once believed that cannabis was not addictive, but more recent research has shown that people may become dependent upon cannabis, which is called "cannabis use disorder" affecting about 10 percent of the cannabis users worldwide.[2] However, most agree that cannabis use disorder does not result in the same extreme levels of behavior apparent with addiction to other drugs of abuse. The features of cannabis use disorder encompass the range of signs and symptoms of other forms of substance use disorders, including tolerance and withdrawal, continuing to use cannabis despite knowing that one has problems caused by cannabis, and recurrent use in situations that might be dangerous like driving a car. Excess cannabis use can take over one's life, and health behaviors and relationships can be harmed. Craving to use cannabis and using for a longer time than planned are common features of this disorder. Essentially, cannabis use disorder manifests in the same way as other substance use disorders, but the difference may be in the magnitude of the severity of the symptoms expressed.[3]

Withdrawal from daily use of cannabis produces symptoms that appear within the first day or two following discontinuations of use, with the effects peaking after 2–6 days and most symptoms resolving in a week or two. These symptoms include, irritability, anxiety, muscle pain, nightmares, insomnia, headache and decreased appetite. These withdrawal symptoms may be relived with oral administration of THC. About 25 percent of patients entering treatment for substance abuse in the United States have a diagnosis of cannabis use disorder.[4]

One reason that cannabis was considered not addictive for many years was the inability of experimental preclinical research to demonstrate the rewarding effects of THC. Unlike drugs known to be addictive, such as opioids and cocaine, laboratory animals did not choose to self-administer THC—this is a hallmark characteristic of an addictive drug. However, this early work evaluated THC administered by injection, rather than the more common human administration of inhalation. Recent preclinical work has shown that even laboratory rats choose to self-administer THC when it is administered by inhalation.[5]

Cannabinoids and Relapse

One of the greatest challenges in the treatment of addiction is relapse to drug use after a period of drug withdrawal. There is a high rate of relapse after abstinence from the addicting drug due to drug-craving. Exposure to environmental stimuli associated with the effects of the drug, exposure to the drug itself, or exposure to stress can precipitate craving and relapse to drug use in abstinent humans. In preclinical research, relapse is assessed by the ability of a cue, drug, or stress to reinstate previously extinguished drug self-administration (lever pressing for administration of a drug) in animal models. It has been shown in preclinical work that CB_1 receptor antagonists (or inverse agonists) like rimonabant can reduce relapses to drug-seeking (craving) for cocaine, heroin, amphetamine and nicotine in rodents by interfering with activation of the mesolimbic dopamine system. There is also considerable evidence suggesting that CB_1 receptor agonists reinstate heroin-seeking, cocaine-seeking, and alcohol-seeking behavior in animal models, suggesting that cannabinoid signaling is involved in drug-craving regardless of the nature of the drug.[6]

The potential of CB_1 antagonists (such as rimonabant) to reduce drug-craving in animal models was of great promise as an antirelapse treatment; however, as we see in chapter 7, rimonabant was removed from the European market as a treatment for obesity because of psychiatric side effects of anxiety and suicidal thoughts. Because rimonabant is a CB_1 antagonist/inverse agonist, the relapse preventing properties, and potentially the adverse side effects, may be mediated by its inverse cannabimimetic effects that are opposite in direction from those produced by cannabinoid receptor agonists.[7] These adverse side effects limit its usefulness as an antirelapse drug. Potentially more recently developed CB_1 neutral antagonists without the adverse inverse agonist side effects may be more promising treatments to prevent relapse.[8]

CBD as a Treatment for Opiate Addiction

Growing preclinical evidence suggests that CBD reduces the potential of drug-associated cues to trigger relapse in animal models. Opiate addiction has been a primary target of the therapeutic potential of CBD treatment. Although CBD does not prevent heroin self-administration ("drug-taking"), it reduces the potential of a heroin-paired cue to trigger "drug-seeking" (craving) for weeks following CBD administration.[9] This may be due to the potential of CBD to interfere with heroin-induced impairments in the dopamine reward system of the brain.[10] As well, CBD reduces opiate withdrawal signs in opiate-dependent animals.[11]

In preclinical research, CBD does not prevent heroin self-administration, but it reduces the potential of a heroin-paired cue to trigger "drug-seeking" (craving) for weeks following CBD administration.[12]

Human clinical evidence has also been provided for the promise of CBD as a treatment for opiate addiction.[13] Based on the success of the preclinical research, a series of clinical studies conducted in humans showed that CBD is safe in humans and did not produce adverse consequences when coadministered with the potent opioid fentanyl.[14] CBD did not modify the subjective effects of fentanyl, but reduced craving and anxiety produced by heroin-associated cues in a human trial.[15] Finally, a double-blind, placebo-controlled RCT explored the effects of acute and short-term CBD administration on craving and anxiety in abstinent women and men with heroin use disorder. A video of heroin-associated cues was found to trigger craving and stress reactivity among the placebo tested group, but CBD (400 mg or 600 mg, Epidiolex, oral) reduced the craving and stress reactivity (heart rate, respiratory rate and salivary cortisol) elicited by the heroin-associated cues, even a week after the CBD treatment.[16] Whereas these findings are encouraging, they do not constitute an efficacy trial for heroin addiction and should not be used to forgo treatments that already exist. Further trials are essential to verify the efficacy of CBD as a potential treatment for opiate addiction

CANNABINOIDS AND EMOTIONAL PROCESSING: ANXIETY, DEPRESSION, AND PSYCHOSIS

Cannabis is commonly used for relaxation. As a result, with the discovery of THC, the earliest experimental investigations in the 1970s were on its effect on anxiety and mood. Activation of the CB_1 receptor is critical for the regulation of the body's systems for signaling stress responses.[1] Healthy individuals given a CB_1 receptor antagonist/inverse agonist (e.g., rimonabant) show increased anxiety, depression, and suicidal ideation. People with major depression and those suffering with Posttraumatic Stress Disorder (PTSD) show reduced levels of endocannabinoids in their circulating blood supply, and there is a negative correlation between endocannabinoid levels and anxiety measures in humans. Recent evidence suggests that treatments that enhance endocannabinoid signaling may be a promising treatment for PTSD.

Cannabinoids and Anxiety/Depression

Cannabis is used frequently by patients with anxiety disorders, especially in times of stress, suggesting that it is used to reduce anxiety. However, long-term use can worsen anxiety and often lead to panic attacks. Laboratory studies with animals suggest that low doses of THC reduce anxiety (anxiolytic effects), but high doses of THC produce anxiety (anxiogenic effects) and stress-induced cortisol release in rats, accompanied by increasing neuronal activity in the limbic system, particularly the amygdala. Cannabis in recent years tends to have higher concentrations of THC, which may be related to the increased reports of panic attacks in humans.[2]

There is evidence from preclinical studies with animals that CBD may have antianxiety effects at low to moderate doses, but no effect on anxiety at higher doses. Unlike THC, CBD does not appear to produce anxiety at high doses—it is either anxiolytic or without effect at higher doses in preclinical studies. Human clinical studies generally confirm that CBD produces an antianxiety effect. In early studies, CBD reversed the anxiety produced by THC in humans. More recently, CBD has been shown to reduce experimentally induced anxiety, especially anxiety elicited by public speaking in healthy human participants and in participants with social anxiety disorder.[3] CBD is safe and well tolerated in humans.

The potential of THC and CBD as treatments for PTSD. Post-traumatic stress disorder (PTSD) is characterized by symptoms of persistent, intrusive recollections, re-experiencing of the original traumatic events (through dreams, nightmares, and dissociative flashbacks), numbing, avoidance, and increased arousal.[4] Nearly 90 percent of patients with PTSD show sleep disturbance.[5] Cannabis use is high among PTSD patients, and cannabis use disorder patients have higher incidence of PTSD, suggesting that this population of patients self-medicate with cannabis. THC (nabilone) reduced nightmares and daytime flashbacks in a subset of veterans.[6]

A behavioral approach to the reduction of PTSD symptoms is that of extinction of fearful memories. Extinction training involves presenting the stimulus that has been associated with the fear-eliciting event (such as shock), but in the absence of the negative event. In laboratory studies with healthy human volunteers, CBD has been shown to enhance the extinction of fear memories; that is, CBD speeds up the reduction of fear to the extinguished stimulus.[7] However, further evidence is required to determine whether CBD may reduce anxiety induced by fearful memories in human PTSD patients themselves. There have been no human clinical trials with PTSD patients being given CBD to reduce the fearful memories.

Cannabis, mood, and depression. Cannabis use is often reported to enhance mood, causing euphoria. Preclinical evidence with animals suggests that THC (at lower doses) and CBD both produce antidepressant effects. Although this suggests that cannabis use might reduce depression, the findings in humans are contradictory.[8] In human self-report questionnaires, results consistently show that people report using cannabis to reduce depression and elevate mood; however, when depressed individuals were given pure oral THC in the laboratory, some patients reported dysphoria (a sense of unease with life), especially those who are naive to the psychoactive effects of cannabis. Without the benefit of the additional cannabis compounds, pure THC may have a different effect than that of cannabis consumption. In fact, humans given only THC show increased anxiety, whereas coadministration of CBD countered its effect.[9] Recent evidence suggests that CBG in cannabis may also reduce anxiety.[10]

Endocannabinoids and anxiety. Since the discovery of the endocannabinoid system, our understanding of the etiology of a "runner's high" (feeling of euphoria and anxiety reduction following exercise) has been revised. It was first believed that endogenous endorphins produce these effects; however, it is now known that both endorphins and AEA levels are elevated in the blood following running. The anxiety reduction after running has been shown to depend on intact CB_1 receptors in forebrain GABAergic neurons.[11] These findings suggest that treatment with endocannabinoids may be effective as anxiety reducers.

Cannabis use is high among PTSD patients, and cannabis use disorder patients have higher incidence of PTSD, suggesting that this population of patients self-medicate with cannabis.

However, AEA is rapidly degraded in the brain by the enzyme, FAAH, and therefore it is not likely that treatment with AEA itself would be useful in suppressing anxiety.

Because AEA and 2-AG are rapidly degraded by the enzymes FAAH and MAGL, respectively, investigators have explored the efficacy of drugs which block these enzymes and thereby prolong the action of natural endocannabinoids by hours as potential treatments for anxiety and depression. Because endocannabinoids are produced only when and where they are needed in the brain in times of stress, their levels will be highest in regions of the brain that regulate emotional behavior. Therefore, inhibition of FAAH or MAGL metabolism in these regions should elevate AEA or 2-AG, respectively, for a matter of hours (up to twenty-four hours) and result in enhanced activation of CB_1 receptors in that region. Several researchers have investigated and are currently investigating the potential of such FAAH or MAGL inhibitors to reduce anxiety and depression in preclinical animal models. This work has revealed that the reduction of anxiety by FAAH inhibition only occurs under highly stressful conditions.[12] Indeed, if mice are exposed to a stressor, then over the next twenty-four hours their anxiety levels are enhanced and AEA levels are decreased; however, pretreatment with a FAAH inhibitor prevented both the anxiety responses and the drop in AEA levels in these mice.[13] Furthermore, elevation of AEA by treatment with FAAH inhibitors promotes extinction selectively of aversive memories (but not rewarding memories), suggesting that FAAH inhibitors may be a potential treatment for PTSD.[14] Individuals genetically deficient in FAAH (with elevated AEA) do not experience anxiety in both preclinical animal and human studies. These findings collectively suggest that treatments that inhibit the degradation of FAAH may be useful in treating stress-related neuropsychiatric disorders. However, there have been no clinical trials with FAAH inhibitors in the treatment of anxiety disorders. There is clearly a need for such studies.

Stress coping and the endocannabinoid system. One of the primary functions of the endocannabinoid system is to regulate stress responses of the body. The body reacts to any potential threats or dangers by neuronal and hormonal responses. The neural responses are immediate (within seconds) followed by hormonal responses over minutes to hours mediated by activation of the hypothalamus-pituitary adrenal (HPA) axis and resulting ultimately with the release of adrenal glucocorticoids. The endocannabinoid system is critical for maintaining homeostasis in this stress system by inhibiting HPA axis activity and reducing the release of stress hormones. Treatments that inhibit FAAH (thereby increasing AEA levels) reduce the hormonal response to stress. AEA is a "gatekeeper" maintaining basal levels of stress hormones (cortisol).[15] Overall, the research suggests a bidirectional effect of stress on the endocannabinoid system—stress exposure first reduces AEA and then increases 2-AG to buffer and constrain the effects of stress on the brain and terminate the stress response.[16]

Cannabis and sleep. Many cannabis users report that they use it to promote sleep, but surprisingly there have been few studies to explore this connection in humans. Sleep disorders are usually treated with sedatives, but side effects, including dependence and tolerance, occur with these drugs. The findings regarding cannabis use and sleep are inconsistent and often lack statistical control for confounding factors. For instance, medicinal users report that cannabis use alleviates sleep problems, but cannabis is also reported as a risk factor for sleep problems.[17] Several studies that included sleep as a secondary treatment outcome, evaluated the assessment of synthetic THC (Marinol, nabilone) or the nabiximols (Sativex CBD/THC) for various conditions (including pain, multiple sclerosis, anorexia, cancer, and HIV). Of a total of 28 studies, 22 of them reported a positive effect of the drug on sleep in the clinical trial; however, it is not clear whether this was the result of improvement of the underlying condition being treated (e.g., pain) that interfered with sleep.[18] PTSD patients did display improvements in sleep time and quality when treated with synthetic THC.[19] As well, medical cannabis and cannabinoids may improve sleep among people living with chronic pain.[20]

A recent review of cannabis and the endocannabinoid system in sleep regulation concluded that cannabis use is not associated with changes in sleep quality; however, THC alone increased daytime sleepiness, whereas CBD alone had no effect on sleep in humans.[21] They concluded that the role of the endocannabinoid system in sleep regulation is critically understudied. An evaluation of the use of CBD in the management of insomnia, the use of a population of patients specifically with a diagnosis of insomnia is lacking in the literature.[22]

Cannabinoids and Psychosis

Cannabis and acute psychosis. It has been known throughout history that cannabis intoxication (more specifically THC intoxication) can elicit acute paranoid psychosis in some users. Jacques-Joseph Moreau reported in 1845 that hashish could induce "acute psychotic reactions, generally lasting but a few hours, but occasionally as long as a week; the reaction seemed related to the amount ingested and its main features included paranoid ideation, illusions, hallucinations, delusions, depersonalization, confusion, restlessness and excitement."[23] Heavy cannabis use can produce psychotic reactions in healthy people and may worsen a preexisting chronic psychotic disorder. Recent evidence suggests that CBD may mitigate some of these effects that are produced by THC.[24] Therefore, CBD may reduce the acute psychotic effect of THC. However, there is also evidence refuting this claim.[25]

Adolescent use of cannabis and later development of chronic psychosis. One of the most controversial issues in debates about cannabis use is the putative link to adolescent use and the later development of psychotic disorders. There is evidence that a greater risk of psychosis occurs when use begins during adolescence than when it begins in adulthood.[26] Insofar as some regions of the brain are still developing during adolescence, the effects of cannabis exposure during that period may affect the developing neural networks. Several observational studies have suggested that regular cannabis use in adolescence is associated with about a twofold increase in the risk of psychosis and regular cannabis use in adolescence has been linked to an earlier age of onset of psychosis; however, it is not clear that these associations are causal in nature.[27]

Retrospective, population-based studies are correlational with the corresponding potential confounds that might account for some of the findings. Such confounds include preexisting differences between adolescents who use cannabis and those who do not; for instance, cannabis users may be more likely to be risk-takers, and this characteristic may also make the users more prone to develop psychotic disorders. That is, a third condition (preexisting personality characteristic, e.g. risk-taking, alcohol or nicotine use, gender, socioeconomic factors) may underly both the tendency to use cannabis in adolescence and a tendency to develop a psychotic disorder. Most people who have used cannabis in adolescence do not develop a psychotic disorder, and most people who develop a psychotic disorder may never have used cannabis; that is, cannabis use is neither necessary nor sufficient to "cause" psychotic disorders.[28] Nonetheless, individuals with a family history of psychosis, individuals with prodromal symptoms, and individuals who have experienced acute episodes of psychosis related to cannabis should be strongly discouraged from using THC-predominant cannabis and psychoactive cannabinoids.[29]

There has been considerable interest in identifying a genetic link between adolescent cannabis use and later propensity to develop schizophrenia. One early potential link was seen among individuals with a specific genotype for the enzyme, COMT (Catechol-0-methyl-transferase) that metabolizes dopamine (DA) in the prefrontal cortex (an area associated with excessive DA activity in schizophrenia). Individuals with the Val/Val genotype for COMT (in contrast to the Met/Met genotype) were reported to be more likely to display psychotic reactions acutely to smoked THC.[30] This led to research to determine whether this Val/Val genotype may serve as a genetic marker to predict individuals likely to develop later schizophrenia in response to cannabis use in adolescence. One study reported that adolescent cannabis users with the Val/Val COMT genotype were more likely to develop psychosis later in life than those without it, but more recent epidemiological studies have not supported these original findings.[31] In fact, Costas et al. showed that schizophrenic patients who were Val/Val COMT genotypes had *lower* rates of lifetime cannabis use than those with the Met/Met COMT genotype.[32] The data are at odds with the earliest report that the Val/Val genotype may serve as a genetic marker of proclivity to develop schizophrenia following cannabis use in adolescence.[33] There have been several other potential genetic markers suggested for the potential of cannabis use to trigger a later psychosis, but most subsequent work was unable to confirm the original results. Therefore, the results supporting the hypothesis that some gene variants may interact with cannabis use to promote the development of schizophrenia are tentative at best. These inconsistent findings limit the likelihood that investigators will find a genetic marker for the link between cannabis use in adolescence and later development of psychosis.

An alternative explanation is that a shared genetic link makes individuals more likely to develop schizophrenia and to use cannabis. That is, those individuals more prone to develop schizophrenia are also more prone to use cannabis and vice versa. A report from a large genome-wide study of 2,082 healthy individuals suggests that the association between schizophrenia and cannabis use may be due in part to shared genetic etiology across common variants. Power et al.[34] found that individuals with an increased genetic predisposition to schizophrenia are both more likely to use cannabis and more likely to use it in greater quantities. Therefore, it is not clear which came first—use of cannabis-triggering schizophrenia or an innate genetic predisposition to schizophrenia triggering use of cannabis; that is, the same genes that increase psychosis risk may also increase the risk of cannabis use. More recent genome-wide association studies (GWAS) have confirmed a shared association between developing a psychotic disorder and using cannabis.[35]

One additional issue has been raised about a purported causal connection between cannabis use and the development of schizophrenia. The incidence of schizophrenia across cultures and worldwide has remained approximately 1 percent of the population. Yet the use of cannabis, especially by adolescents has dramatically increased beginning in Western cultures in the 1960s and 1970s. In the United States and Canada, approximately 25–30 percent of high school–aged young adults have used cannabis and about 5 percent use it daily. However, there has not been a corresponding increase in the incidence of schizophrenia—it remains approximately 1 percent of the population in these countries. It remains approximately 1 percent in countries in which there is a high rate of cannabis use in adolescence (the United States and Canada) as well as countries in which the rate of cannabis use in adolescence is much lower (e.g., Scandinavian countries, Greece, Italy, and China). The causal argument has difficulty explaining these observations.

A consensus paper has recently been published authored by several of the major researchers investigating the potential links between cannabis use and psychosis.[36] These authors conclude, "Converging lines of evidence suggest that exposure to cannabis increases the risk for psychoses ranging from transient psychotic states to chronic recurrent psychosis. The greater the dose, and the earlier the age of exposure, the greater the risk. For some psychosis outcomes, the evidence supports some of the criteria of causality. However, alternate explanations including reverse causality and confounders cannot be conclusively excluded. Furthermore, cannabis is neither necessary nor sufficient to cause psychosis. More likely it is one of the multiple causal components. Given that exposure to cannabis and cannabinoids is modifiable, delaying or eliminating exposure to cannabis or cannabinoids could potentially impact the rates of psychosis related to cannabis, especially in those who are at high risk for developing the disorder."

CBD as antipsychotic. There is growing interest in the possibility that CBD may be a potential antipsychotic agent. The prevalence of cannabis-linked acute psychosis has been reported to be lower when street cannabis contains a higher proportion of CBD and in healthy humans CBD may reverse THC-induced psychotic symptoms. CBD (up to 800 mg/day) has been compared in a controlled clinical trial with the standard antipsychotic drug, amisulpride.[37] After 4 weeks, both groups showed significant improvements from baseline; however, CBD showed a superior safety profile, with fewer adverse side effects. Positive results were also reported in the treatment of schizophrenia with CBD (1 gm/day for 6 weeks) in a double-blind placebo control trial with both groups continuing to receive their prescribed antipsychotic medicine.[38] A third positive double-blind trial found that high-dose CBD reduced distress in clinical high risk psychosis patients.[39] However, Boggs et al. using a lower dose of CBD (600 mg) failed to show improvement of symptoms with CBD.[40] CBD has also been demonstrated to be effective in treatment of psychotic symptoms as well as motor symptoms in patients with Parkinson's disease.[41] These results suggest that CBD, which repeatedly has been shown to lack toxicity and to produce few adverse side effects in normal healthy participants (in contrast with the usual antipsychotic drugs), may ameliorate psychotic symptoms in schizophrenic patients, albeit at very high doses.

CANNABINOIDS: COGNITION AND BRAIN DEVELOPMENT

Cognitive abilities involve acquiring, storing, and later retrieving new information. THC can produce cognitive disruptions in humans and animals, affecting short-term memory more profoundly than retrieval of long-term memory. Disturbances in the ability to hold information in short-term memory (e.g., a telephone number) for brief periods are particularly impaired. The disruption of short-term memory during a state of intoxication prevents the transfer of this new information into long-term memories; that is, THC interferes with consolidation of new memories.

Preclinical Evaluation of Cannabis Effects on Cognition

Considerable preclinical animal research has evaluated how CB_1 agonists affect learning and memory. Some of these studies involve administration of THC and other global CB_1 agonists; others involve manipulations of the endocannabinoid system by administration of FAAH inhibitors (which elevate AEA) and MAGL inhibitors (which elevate 2-AG) when and where they are produced. Consistent with the human literature, the effect of CB_1 agonists selectively disrupts short-term or working memory, but not long-term or reference memory. The direct effect of CB_1 agonists also reduces the consolidation of short-term memories into long-term memory.

The decrement in short-term memory caused by cannabinoids is mediated in a large part by their action at the hippocampus (although other brain regions are also involved), which has a high density of CB_1 receptors. The detrimental effects of CB_1 agonists on short-term memory parallel the effects of hippocampal lesions.[1]

Memory formation in the hippocampus is believed to be mediated by long-term changes in synaptic plasticity, called *long-term potentiation* (LTP). One of the most interesting characteristics of LTP is that it causes long-term strengthening of the synapses between two neurons that are activated simultaneously. The excitatory neurotransmitter, glutamate, plays a major role in synaptic plasticity resulting in the formation of new memories. LTP relies on the activation of NMDA (N-methyl-D-aspartate) glutamate receptors. CB_1 agonists impair consolidation of memories by suppressing the release of glutamate in the hippocampus and thereby preventing LTP. The long-term suppression of communication across two synapses, called *long-term depression* (LTD), is the opposite of LTP serving as a mechanism for memory impairment by cannabis. LTD is characterized by a decrease in postsynaptic strength. Retrograde signaling by endocannabinoids results in suppression of neurotransmitter release at both excitatory (glutamatergic) and inhibitory (GABAergic) synapses in the hippocampus in a short-term manner, suppressing LTP, and in a long-term manner, producing LTD, one of the best examples of presynaptic forms of long-term plasticity. Recent evidence indicates that presynaptic activity coincident with CB_1-receptor activation and NMDA-receptor activation is required for some forms of endocannabinoid LTD resulting in memory impairment.[2]

Cannabinoid receptors localized to different brain regions modulate distinct learning and memory processes, and so the role of endocannabinoids in other regions may be different from their role in the hippocampus. For instance, infusion of a CB_1 agonist into the basolateral amygdala in rodents enhanced consolidation of inhibitory avoidance learning (learning to withhold a response), and a CB_1 antagonist into this region interfered with avoidance learning.[3] The differential effects of CB_1 agonists and antagonists on different brain regions may account for the different findings reported after systemic and localized administration of cannabinoid agonists.

Cannabis and Neurocognition in Humans

Acute effects of cannabis on cognition. In humans as well as other animals, THC produces a transient impairment in short-term memory and in consolidation of short-term memory into long-term memories; however, it does not produce an impairment in the retrieval of previously consolidated memories. These effects of THC do not appear to extend beyond the period of intoxication. A controlled study revealed no residual cognitive impairment twenty-four hours following acute exposure to THC.[4] There is some evidence that CBD may protect against the memory deficits produced by THC.[5]

Chronic effects of cannabis use on cognition. It is not clear if long-term chronic exposure to THC produces lasting cognitive deficits that persist when the individual is not experiencing the acute effect of exposure. A major systematic review concluded that the literature is unclear about the long-term effects of cannabis use on cognition in terms of the amount of cannabis used and means by which cognition was assessed.[6] Among chronic users, the effects may be attributable to chronic long-term use or to residual or cumulative acute effects resulting from the buildup of THC in tissue. Confounding factors that may attribute cognitive impairment to cannabis include preexisting differences in cognitive ability between users and nonusers and other substance abuse. Very few large sample studies controlled for cognitive ability that was assessed before initiation of cannabis use.

A high-profile study suggested that heavy cannabis use in adolescence was associated with a decline of approximately eight IQ points from childhood to early adulthood—this has been coined "the Dunedin study" conducted in New Zealand.[7] This prospective study followed participants from birth (1972 or 1973) to age thirty-eight, therefore a change from precannabis use to postcannabis use is evident for each participant. They found that persistent heavy cannabis use (with a diagnosis of cannabis use disorder in at least three sampling periods) was associated with neuropsychological decline across domains of cognitive functioning, even controlling for years of education. The decline was most significant among those who had been heavy users in adolescence and early adulthood. Late onset (in adulthood) heavy users among the sample did not show a persistent deficit after quitting cannabis, but early onset heavy users maintained their deficit after abstinence in adulthood. This suggested that heavy cannabis use in adolescence may produce persistent cognitive deficits even after one abstains from cannabis use. On the other hand, two additional large prospective studies have since failed to replicate the findings of the Dunedin study, one in the United Kingdom and one in the United States.[8] In both studies, it was found that cumulative cannabis use was not associated with a lower IQ relative to nonusing controls when IQ measures were compared against those taken before the teen years and when alcohol and tobacco use were removed as potential confounders. Although these recent studies are limited by participants reporting fewer cannabis exposures than the Dunedin study, there does not appear to be support for a causal link between cannabis use and long-term change in IQ.[9]

Does chronic use of cannabis produce long-term impairment of cognitive abilities in abstinence from cannabis use? The literature is fraught with contradicting answers to this question. All such studies rely on retrospective self-reporting of cannabis use, and preexisting cognitive and emotional differences between cannabis users and non-users as well as polydrug use make interpretation of the literature problematic.[10] When the analyses are limited to studies of cannabis users having had at least one month of abstinence, there is no discernible difference between cannabis users and nonusers in performance on neurological tests; therefore, cognitive functions may recover with prolonged abstinence.[11] A number of studies of long-term effects in abstinence suggest that cognitive impairments do not persist beyond 4–6 weeks after abstinence.[12] Cortical CB_1 receptors show downregulation (decrease in density) among chronic cannabis users that is correlated with years of use, but after around 4 weeks of continuously monitored abstinence their CB_1 receptor density returned to control levels.[13] This effect has been reported even after just two days of abstinence and similar findings have been reported from preclinical animal studies.[14]

Effects of adolescent cannabis exposure on cognition and brain development. Adolescence is a period of delayed frontocortical maturation; the thinking brain becomes more efficient and isn't fully developed until the early twenties. As stated by Robert Sapolsky in his book *Behave*, "Because it is the last to mature, by definition the frontal cortex is the brain region least constrained by genes and most sculpted by experience."[15] Such experience includes exposure to cannabis! Does exposure to cannabis in adolescence produce greater memory impairments than exposure to cannabis in adulthood? In preclinical research, adolescent rats exposed to THC showed increased CB_1 gene expression compared to adult THC exposed rats, as well as disruption of the excitatory-inhibitory balance of GABA and glutamate in the hippocampus compared with adult THC exposed rats.[16] The extent to which these deficits persist following an adequate period of abstinence remains unclear.

Cannabis use and the potency of THC in cannabis has increased in the recent years. Because it is known that THC produces at least acute effects on short-term memory processes, it has been suggested that long-term chronic use of high-potency cannabis may produce permanent neural changes affecting cognitive processes, which may be most profound in adolescence. However, whether brain changes result from long-term cannabis use is equivocal. Some studies report decreased volume of subcortical regions in long-term cannabis users; others report increased volumes in the same or different regions in chronic cannabis users.[17] Indeed, when the sizes of effects are averaged across all studies, Weiland et al.[18] report that there is no statistical effect.

Most human studies regarding neurocognitive effects of cannabis use employ a cross-sectional approach that is unable to disentangle the nature of the relationship between adolescent cannabis use and neurocognitive alterations. That is, it is unclear whether adolescent cannabis use is a cause or consequence of these neurobiological alterations or if their association is linked to an underlying predisposition that increases the likelihood of both adolescent cannabis use and certain neurocognitive outcomes.[19] The literature needs larger samples, twin studies and longitudinal studies to disentangle this potential confound.

It is challenging to draw conclusions from the human cannabis literature regarding the effects on cognition because investigators use widely differing methods, differing participant-selection techniques, differing cannabis strains/doses/administration techniques, sample sizes are often small, and, most notably, other drug use is common. Predisposition to substance use in general may confer greater vulnerability to cannabis related cognitive effects.

That is, people who misuse substances typically differ from people who do not on important characteristics that may influence both cannabis use and brain outcomes. Despite significant public health implications, there remains a lack of evidence separating sources of causality related to cannabis use on the emerging adult brain.

Twin studies provide ideal genetic and shared environmental controls to evaluate causal effects of cannabis and alcohol exposure effects on the development of cortical regions implicated in cognitive control. Such studies are unconfounded by familial influences that is not possible in cross-sectional or longitudinal studies of genetically unrelated individuals. In one recent study of 436, 24-year-old twins used a co-twin control analysis to attempt to unravel some of these confounds.[20] Twins were evaluated for their intake of cannabis and alcohol. Outcomes of the lesser drug-using twin provided a close approximation of the familial expected outcome of the heavier drug-using twin. This comparison can be used to address a causal relationship between cannabis and alcohol use in emerging adulthood and the cortical thickness in regions mediating cognition control. The results revealed that heavy alcohol use, but not cannabis use, was a strong predictor of cortical gray matter decreases in thickness, supporting previous suggestions that comorbid alcohol use may account for previous findings of a link between cannabis use and cortical thickness changes in adolescents. However, this finding does not rule out the possible effects of cannabis in adolescence on subcortical development. This study suggests that alcohol abuse may account for some of the structural brain changes evident in long-term cannabis users reported in earlier studies that did not take this into account.

Larger-scale longitudinal studies currently underway (i.e., the ABCD-Adolescent Brain and Cognitive Development Study at the National Institutes of Health) may provide a better understanding of the effects of chronic exposure to marijuana on cognitive function independent of preexisting differences or comorbidity with alcohol. Therefore, our current understanding of the effects of cannabinoids on various processes involved in learning and memory rely heavily on animal models, which provide insights into the role of the endocannabinoid system in the physiology of learning and memory.

CANNABINOIDS: FEEDING, BODY WEIGHT, AND APPETITE

"The munchies" are one of the most well-known effects of cannabis use—stimulation of appetite particularly for palatable foods. Shortly after the discovery that THC was the primary psychoactive compound in cannabis, Leo Hollister reported that a single oral dose of cannabis increased the intake of milkshakes in healthy volunteers.[1] Others verified this effect with smoked cannabis.

Cannabis has the therapeutic effect of stimulating appetite in the treatment of cachexia, a chronic wasting disorder associated with loss of adipose tissue and lean body mass seen in patients with cancer, acquired immunodeficiency disorder (AIDS), and anorexia nervosa. Synthetic THC (dronabinol) is used in the clinic to combat the reduction in appetite in such patients with mixed results. Some studies have found that dronabinol (synthetic THC, 2.5 mg twice a day) enhanced appetite and body weight

"The munchies" are one of the most well-known effects of cannabis use—stimulation of appetite particularly for palatable foods.

in AIDS patients suffering from anorexia; whereas other studies with both AIDS patients and patients with cancer-related cachexia report no effect with a similar dose and regime. Orally administered pure THC may not be the optimal treatment for these disorders. One alternative is the use of FAAH inhibitors (elevating AEA centrally, but lacking psychoactivity), which have been shown to enhance motivation for food and promote energy storage in preclinical animal studies. As we learn more about the effects of cannabinoids and direct manipulations of the endocannabinoid system on appetite, better treatments may be developed.

Endocannabinoids and Regulation of Body Weight

One of the most well-studied effects of the endocannabinoid system is its role in the maintenance of energy balance. In the brain, the endocannabinoid system interacts with reward pathways in the mesolimbic dopamine system and the hypothalamus that regulate levels of hormones that enhance feeding. Endocannabinoid levels in the hypothalamus and reward pathways are highest during food deprivation, leading to food-seeking behaviors. The neural circuitry of the hypothalamus uses this information to regulate caloric intake, energy consumption, and peripheral lipid and glucose metabolism. Administration of AEA into the hypothalamus induces eating, and endocannabinoid levels in the hypothalamus vary as a function of nutritional status. The action of endocannabinoids in the hypothalamus regulates the levels of several biochemical compounds that control feeding and body weight, including melanocortin (which reduces feeding) and neuropeptide Y (which stimulates feeding). Within the hypothalamus, modulation of the expression of several pro- and antifeeding hormones by the endocannabinoid system is counterbalanced by the opposite actions mediated by the adipose-derived hormone, leptin.

Leptin, a key player in the regulation of appetite and hunger, originates in adipose tissue and affects several appetite-related factors in the hypothalamus. Endocannabinoid levels in the hypothalamus inversely correlate with leptin plasma levels. Leptin administration (which exerts an anorectic action) suppresses hypothalamic endocannabinoid levels in healthy animals, but in the hypothalamus of obese and hyperphagic (overeating) rodents lacking leptin, endocannabinoid levels are significantly increased. Functional CB_1 receptor signaling in the hypothalamus is required for leptin to exert its suppressive effects on food intake; selective genetic knockout of CB_1 receptors in the hypothalamus of mice abolished the inhibition of food intake by leptin. This inverse relationship also affects the activity of the brain reward system; obese rats with defective leptin signaling show increased CB_1 expression and binding activity in brain reward structure.

The endocannabinoid system also regulates food intake and energy balance by its action on the vagus nerve that carries information between the gut and the brainstem. Within the gut, cholecystokinin (CCK) acts to suppress feeding (a satiety signal) and ghrelin acts to stimulate feeding. After feeding CCK decreases, and after fasting it increases CB_1 receptor expression in vagal afferents. The decrease in CB_1 receptor expression following feeding is prevented by administration of ghrelin. Therefore, ghrelin opposes the action of CCK on CB_1 receptor expression on vagal afferents from the gut to the brainstem. Ghrelin has also been shown to elevate hypothalamic endocannabinoid content.

Only a single clinical study has investigated the regulation of endocannabinoid levels in humans eating food for pleasure. Plasma levels of 2-AG and ghrelin (but not AEA) were elevated in normal-weight healthy humans following hedonic eating. In mice, 2-AG levels in the hypothalamus increased in response to high-fat diet. Highly palatable foods may evoke alterations in the central nervous system that is similar to that of drugs of abuse through the regulation of common neurobiological substrates. Overconsumption of palatable foods is accompanied by the stimulation of the brain dopaminergic and opioid reward systems.

Endocannabinoid System in Eating Disorders and Obesity

The endocannabinoid system is a modulator of both homeostatic and hedonic aspects of feeding; therefore, dysfunctions in this system may lead to eating disorders. Anorexia nervosa and bulimia nervosa are characterized by abnormal eating behaviors resulting in severe food restriction with a dramatic loss of body weight in anorexia nervosa and episodes of binge eating and vomiting without significant changes in body weight with bulimia nervosa. As well, the *Diagnostic and Statistical Manual of Mental Disorders-IV* (*DSM-IV*) includes a category of binge-eating disorder, which is characterized by binge eating but without compensatory vomiting resulting in obesity. Such eating disorders clearly involve both psychosocial and biological factors.

Among other biological factors, dysfunctions in the endocannabinoid regulatory system may be involved in the pathophysiology of eating disorders. Comparison of the blood levels of AEA and 2-AG of women with one of these eating disorders with those of healthy controls revealed increased AEA (but not 2-AG) in patients with anorexia nervosa, but not bulimia nervosa or binge-eating disorder.[2] As well, anorexic women showed decreased circulating leptin levels, and binge-eating disorder women showed increased leptin levels compared with healthy controls. In both healthy controls and anorexia nervosa women, the higher the blood AEA levels the lower the plasma leptin concentrations. This suggested that the decreased leptin signaling of underweight anorexia nervosa patients could be involved in the increase of AEA levels. It is possible that dysregulated endocannabinoid tone of anorexia nervosa patients may represent an adaptive response aimed at maintaining energy balance by potentiating internal food-seeking signals, and hence stimulating food ingestion. However, human studies using dronabinol (oral THC) have largely reported no improvement of anorexia symptoms.

Endocannabinoids tend to be elevated in obesity. Genetically obese mice were found to have hypothalamic endocannabinoid levels higher than in lean animals. Obesity is positively related to either overproduction of endocannabinoids or increased expression of the CB_1 receptor in tissues involved in energy homeostasis in both animal and human studies. The endocannabinoid system has been reported to be dysregulated in overweight obese women with a binge-eating disorder and in obese postmenopausal women. Patients with elevated abdominal fat levels have higher levels of 2-AG in adipose tissue samples compared with patients with subcutaneous fat or lean controls. As well, AEA levels are elevated in the saliva of obese subjects and directly correlate with Body Mass Index, waist circumference, and fasting insulin. Salivary AEA levels may serve

as a useful biomarker for obesity, because among these participants, body weight loss after a 12-week program, decreased salivary AEA levels.

The Rimonabant Story

Because the endocannabinoid system is overactivated in overweight and obese subjects, current therapeutic targets for obesity aim to restore a normal endocannabinoid tone by drugs that interfere with endocannabinoid signaling, such as the CB_1 inverse agonist/antagonist, rimonabant, also known as Acomplia® developed by Sanofi-Avantis.

Laboratory studies in rats and mice provided evidence for the hypothesis that antagonism of the endocannabinoid system with rimonabant may be a promising therapy for diet-induced obesity. In preclinical animal studies, rimonabant reduced food intake and body weight gain. These effects were mediated by the action of rimonabant on the CB_1 receptor, because genetically modified mice lacking CB_1 receptors did not show the weight loss. In preclinical rodent studies, rimonabant also reduced the rewarding effects of foods and drugs and counteracted the increase of extracellular DA release induced by highly palatable foods. THC increased hedonic reactivity to sucrose and CB_1 antagonism did the opposite. Thus, the CB_1 receptor is an important component of the neural substrate that mediates the reinforcing and motivational properties of a highly palatable food.

Several human clinical trials verified the initial anti-obesity preclinical experimental findings with rimonabant; these trials were grouped in the four rimonabant in obesity (RIO) studies.[3] Rimonabant was approved for human clinical use and introduced into the European market as an antiobesity agent in several countries, not including the United States. In the United States, the FDA asked for further evidence regarding safety before approving its marketing.

Unfortunately, following its use as an antiobesity treatment in the European market, several unwanted psychiatric side effects of depression and suicidal ideation became evident among those prescribed rimonabant. A subsequent meta-analysis of the human clinical trial evidence in the four RIO studies suggested that patients treated with rimonabant were 2.5–3 times more likely to experience psychiatric adverse effects, such as anxiety and depression, than patients receiving placebo. The European Union Committee for Medicinal Products for Human Use concluded that rimonabant doubled the risk for psychiatric disorders, and the European Medicines Agency (EMA) suspended the license of the drug. Sanofi-Aventis withdrew the product from the worldwide market, and the clinical development not only of rimonabant but also other CB_1 antagonists being developed by other companies was halted.

New Potential Treatments for Obesity Exploiting the Endocannabinoid System

Rimonabant had high promise for reducing obesity but had unacceptable adverse CNS side effects. This has led several research groups to find compounds with the benefits of rimonabant but without the adverse anxiety and depression side effects in the brain. Recent preclinical investigations have focused on the evaluation of CB_1 antagonists/inverse agonists that do not penetrate the brain as potential antiobesity treatments. Such treatments may be devoid of the adverse psychiatric side effects produced by brain-penetrate CB_1 antagonists/inverse agonists, such as rimonabant. Human peripheral adipose tissue has been shown to possess a fully functional endocannabinoid system, and high-fat diet induces an increase of AEA in the liver due to reduced degradation by FAAH. As well, 2-AG levels are elevated in the visceral fat of obese patients. These findings suggested that the suppression of food intake by rimonabant may be mediated by its action on peripheral CB_1 receptors rather than central receptors. Peripherally restricted CB_1 receptor antagonist/inverse agonists have been demonstrated to produce equivalent reduction in food intake, body weight, and adiposity as with rimonabant without central action.[4] These results are particularly exciting, because they may lead to new treatments for obesity without the possibility of psychiatric side effects that prevented the use of rimonabant as a treatment.

Another approach to reducing the impact of endo-cannabinoids on feeding is to modify their synthesis. For instance, decreasing the synthesis of 2-AG levels by administration of an inhibitor of the enzyme responsible for its synthesis (DAGL) inhibited palatable or high-fat diet intake in mice.[5] As well, mice genetically altered to over-express, MAGL (the enzyme that degrades 2-AG), showed increased energy expenditure and decreased weight gain in a high-fat diet regimen.[6] Another strategy that is being explored is related to the allosteric modulation of CB_1 receptor by changing the binding of the endocannabinoid. A novel allosteric antagonist of the CB_1 receptor reduces that impact of AEA or 2-AG when it binds to the orthosteric receptor site. This allosteric antagonist inhibited appetite and produced weight loss in rats.[7] As well, there is some evidence that CB_2 antagonism has the potential to manage obesity-associated metabolic disorders; this is important because CB_2 agonists do not produce central effects.[8]

What About the Other Cannabinoids in Cannabis?

The psychoactive side effects of THC limit its usefulness as an appetite-stimulating agent in treating anorexia or cachexia—what about other cannabinoids in cannabis that are not psychoactive?

THCV. Given the withdrawal from the market of CB_1 inverse agonists, such as rimonabant because of unwanted psychiatric side effects, safer alternatives are necessary. THCV is a CB_1 antagonist without inverse agonist effects that reduces food intake in rats.[9] It is possible that the lack of inverse agonism of the CB_1 receptor with THCV may render it more tolerable. Therefore, it will be interesting to determine whether THCV may be developed as a potential appetite reducing, antiobesity drug.

CBD. The potential of CBD to modify food intake is controversial, with some studies showing a reduction in feeding and others showing no effect across several differing methodologies. There has been no clinical trial with CBD alone as a treatment for feeding disorders. Given that CBD has been shown to ameliorate some unwanted effects of THC (anxiety, learning deficits), it may be worthwhile to evaluate combined doses of CBD and THC in treatment of obesity.

CANNABINOIDS: NAUSEA AND VOMITING

Nausea and vomiting are important reflexes to protect us against contaminated or potentially harmful foods. These protective reflexes are commonly activated as side effects of medications, most notably cancer chemotherapy agents. Current antiemetic therapies (i.e., ondansetron) are highly effective in reducing chemotherapy-induced vomiting, but they are only weakly effective in reducing chemotherapy-induced nausea. Because nausea is so poorly understood, effective treatments are very limited. Yet nausea is reported to be one of the most debilitating human sensations. Therefore, new treatments for nausea are urgently required. Cancer patients receiving chemotherapy treatment also often experience delayed (24–48 hours) nausea and vomiting, which is also not well controlled by available antiemetic treatments. Finally, when the initial acute emetic episode is not prevented, patients often experience anticipatory nausea and vomiting (classically conditioned responses) upon returning to the clinic. When anticipatory nausea or vomiting occur, there are no effective treatments, other than nonselective sedatives; that is, ondansetron does not work.

THC as an Antiemetic

A report by the National Academy of Sciences concluded that, based on the available studies, there was sufficient evidence for the effectiveness of oral cannabinoids for the treatment of chemotherapy-induced nausea and vomiting. Indeed, survey data from those using medical cannabis to manage their chemotherapy-induced nausea vomiting suggest that these patients are experiencing benefits. The first recognized medical use of THC in modern medical history was for the treatment of nausea and vomiting. In the late 1970s and early 1980s, ineffective treatment of chemotherapy-induced nausea and vomiting prompted oncologists to investigate the antiemetic properties of cannabinoids, before the discovery of the antiemetic effects of 5-hydroxytryptamine-3 (5-HT$_3$) receptor antagonists, such as ondansetron. The first synthetic CB$_1$ agonist, nabilone (Cesamet), was specifically licensed for the suppression of nausea and vomiting produced by chemotherapy treatment. A few years later, synthetic THC, dronabinol,

The first recognized medical use of THC in modern medical history was for the treatment of nausea and vomiting.

entered the clinic as Marinol in 1985 as an antiemetic and in 1992 as an appetite stimulant. In the early studies, several clinical trials compared the effectiveness of THC with placebo or other antiemetic drugs that were available at the time. The synthetic CB_1 agonists were found to be either equivalent to or more effective than other antiemetic treatments available at the time.[1]

Interest in the use of cannabinoids for the treatment of acute vomiting waned in the late 1980s with the development of 5-hydroxytryptamine–3 (5-HT$_3$) receptor antagonists, such as ondansetron. These agents were found to be highly effective in suppressing acute vomiting in preclinical animal models. In clinical trials with humans, treatment with 5-HT$_3$ antagonists, often combined with the corticosteroid dexamethasone during the first chemotherapy treatment reduced the incidence of acute vomiting by approximately 70 percent. Although these 5-HT$_3$ antagonists are highly effective in reducing acute vomiting, they are much less effective in suppressing acute nausea or in suppressing delayed nausea and vomiting occurring 24–48 hours later. They were also ineffective in suppressing anticipatory nausea and vomiting upon returning to the clinic in which the treatment occurred. More recently, NK$_1$ receptor antagonists (e.g., aprepitant) have been developed that not only decrease acute vomiting but also decrease delayed vomiting induced by cisplatin-based chemotherapy; however, these compounds alone and in combination with 5-HT$_3$ antagonist treatment are also much less effective in reducing nausea.

THC as a Treatment for Nausea

Cannabis-based medicines may be particularly effective in treating the more difficult to control symptoms of nausea and delayed nausea and vomiting. A well acclaimed clinical trial evaluated the effectiveness of Δ^8-THC, a close but less psychoactive relative of Δ^9-THC, to reduce nausea and vomiting in children receiving chemotherapy treatment. The children were administered drops of Δ^8-THC two hours prior to the chemotherapy treatment and every six hours thereafter for 24 hours. The only side effects reported were slight irritability in two of the youngest children (3.5 and 4 years old); yet, both acute and delayed nausea and vomiting were very well controlled. This small-scale study (without a placebo control) provides suggestive evidence that these cannabinoids may be particularly well suited for treating nausea and vomiting in children being treated for cancer.

There has been only one double-blind, placebo-controlled clinical trial which has compared the anti-emetic/antinausea effect of cannabinoids with those of 5-HT$_3$ antagonists, and none has compared cannabinoids with the NK$_1$ antagonist, aprepitant.[2] Meiri et al. found that the efficacy of synthetic THC (dronabinol) was comparable to that of the 5-HT$_3$ antagonist, ondansetron, in controlling acute and delayed vomiting; *however, the dronabinol group reported the lowest nausea intensity on a visual analogue scale*.[3] The dose of dronabinol used in this study was at least 50 percent lower than in previous studies conducted in the 1980s resulting in a low incidence of CNS-related adverse effects, which did not differ from the incidence in the ondansetron treated group. The results of this study suggest that THC may be more effective than 5-HT$_3$ antagonists in treating nausea, in particular.

Combined THC/CBD (Nabiximols) as a Treatment for Nausea and Vomiting

Nabiximols (Sativex, 2.7 mg THC: 2.5 mg cannabidiol/spray) was evaluated in a phase 2 clinical trial to determine whether it would enhance the effectiveness of standard antiemetic treatment for chemotherapy-induced nausea and vomiting. Nabiximols or placebo was administered to patients also receiving a 5-HT$_3$ antagonist and a corticosteroid. Nabiximols facilitated relief of delayed nausea and vomiting among patients receiving moderately emetogenic cancer chemotherapy. Approximately half (57 percent) of the nabiximols-treated patients experienced no delayed nausea, and an even greater number (71 percent) experienced no delayed emesis compared with patients receiving placebo. However, given the known antinausea/antiemetic effects of CBD demonstrated in animal models, it is unclear how much of the antinausea effects of nabiximols are attributable to THC and how much to CBD; nonetheless, this study demonstrates the therapeutic potential of combining these cannabinoids in treating delayed nausea and vomiting.[4]

A phase 2 multicenter, randomized, double-blind, placebo-controlled trial evaluated oral nabiximols (2.5 mg THC + 2.5 mg CBD capsule, three times daily, from days 1 to 5, and 1 cycle of matching placebo in a crossover design) for the prevention of chemotherapy-induced nausea and vomiting. The addition of the cannabis extract to the antiemetic regimen reduced nausea and vomiting. A recent randomized, double-blinded, crossover, and placebo-controlled trial examined the effects of nabiximols oil (1:1, THC:CBD) in gynecologic cancer patients, with doses ranging between 0.25 and 0.4 mg. In the first two days following chemotherapy treatment, those patients receiving the nabiximols oil (plus standard antiemetic agents) reported less nausea in comparison to those who received only standard antiemetic. There were no differences in nausea score between groups on days 3 and 4 postchemotherapy, but on day 5 the THC:CBD oil patients reported less nausea than controls. These findings suggest that THC:CBD oil may help to alleviate acute and delayed nausea.

Animal models show that CBD alone may be an effective treatment for nausea and vomiting.[5] However, there have been no human RCTs to verify the efficacy of CBD alone in reducing nausea and vomiting in humans. Only one case report has been published with CBD alone in two patients undergoing chemotherapy and radiation; when taking CBD during their cancer treatment period, two male patients with gliomas reported little nausea.[6] Future randomized placebo-controlled trials with a larger number of patients are needed to properly examine CBD's effects on chemotherapy-induced nausea and vomiting.

Endocannabinoids and Nausea and Vomiting

The administration of cannabinoids systemically (oral, smoking, and sublingual) produces global activation of the CB_1 receptors. On the other hand, given that endocannabinoids are synthesized on demand, manipulations that target endocannabinoid hydrolysis result in a much more localized increase in levels of 2-AG and AEA compared with systemic administration of cannabinoid receptor agonists. Therefore, a localized increase in endocannabinoid levels in the brain regions responsible for nausea and vomiting is less likely to produce unwanted side effects attributable to a global effect and is, thus, preferable for selectively reducing nausea and vomiting.

Considerable preclinical evidence indicates that treatments that boost the endocannabinoid system reduce nausea and vomiting in animal models.[7] The effects of the exogenous delivery of these endocannabinoids are short-lived, because AEA is rapidly degraded by the enzyme FAAH and 2-AG is rapidly degraded by enzyme MAGL. FAAH inhibitors and MAGL inhibitors (which elevate AEA and 2-AG respectively for up to twenty-four hours) suppress both acute and anticipatory nausea in rats and vomiting in shrews.[8] Given that conventional antiemetic treatments are completely ineffective in treating anticipatory nausea and relatively ineffective in treating acute nausea, there is great promise for treatments that boost the endocannabinoid system in the treatment of chemotherapy-induced nausea and vomiting. However, there have been no human RCTs on the potential of FAAH inhibitors or MAGL inhibitors to reduce nausea and vomiting.

There have been recent advancements in understanding the regulation of nausea by the endocannabinoid system of the brain. Brain imaging studies in humans has identified the insular cortex as a region regulating the sensation of nausea.[9] In the insular cortex, the neurochemical trigger producing the experience of nausea appears to be serotonin, which is released by afferents arising from the midbrain region of the dorsal raphe nucleus.[10] This nausea experience and the release of serotonin are prevented by treatments that elevate 2-AG in the insular cortex.[11] That is, the endocannabinoid, 2-AG, acts to regulate nausea in the insular cortex by reducing the release of nausea-inducing serotonin by its action on the CB_1 receptor on serotonergic neurons. This reduction in serotonin may reduce the experience of nausea.

Cannabis Hyperemesis Disorder

Despite the evidence for the antiemetic potential of cannabinoids in humans, it is worthwhile to note that chronic marijuana use has also been reported to produce the sensation of nausea and vomiting. It appears that low doses of THC reduce nausea and vomiting, but high doses may produce nausea and vomiting in some individuals (another instance of the biphasic effects of cannabinoids). This paradoxical effect of cannabis, known as cannabinoid hyperemesis syndrome, has been documented in numerous case reports in recent years. Cannabis hyperemesis disorder is characterized by severe cycles of nausea, vomiting, and abdominal pain. Recovery is produced by abstinence from cannabis, although paradoxically hot showers tend to reduce the symptoms.

Although the cause of hyperemesis is not known, it can be speculated that changes in CB_1 receptor expression following years of exposure to THC in chronic cannabis users may play a role in this paradoxical effect. Chronic exposure to high doses of THC may result in dysregulation of the endocannabinoid system. When CB_1 receptor signaling is high, as occurs in the presence of high doses of agonists or prolonged exposure, receptors become desensitized. Such receptor adaptations result in loss of inhibitory control by the action of THC on the CB_1 receptor which dampens endocannabinoid signaling. This mechanism may underlie the increase in nausea and vomiting as well as the increase in stress and anxiety produced by chronic high-dose THC exposure.[12]

In preclinical research with rats and shrews, DeVuono et al. have shown that although low doses of THC reduce nausea and vomiting, high doses produce nausea and vomiting reactions.[13] Furthermore, the effect of exposure to high doses of THC is accompanied by downregulation of CB_1 receptors and stress responses in these animals. CBD reversed these emetic effects of high doses of THC in these animal models.[14] Further research is undoubtedly needed to fully understand the mechanism(s) underlying this peculiar effect of cannabinoids that may lead to increases in nausea and vomiting following chronic consumption of high doses of THC among human users. Nevertheless, these cases highlight the dynamic nature played by the endocannabinoid system in regulating nausea and vomiting, and they suggest a possible consequence of endocannabinoid system dysregulation, which may lead to undesirable results such as an increased sensation of nausea and emesis.

CANNABINOIDS AND PAIN

The most common medical use of cannabis is for the treatment of chronic pain, which afflicts nearly 20 percent of the world's population. Chronic pain includes nociceptive pain and neuropathic pain. Nociceptive pain is the response of pain receptors (specialized free nerve endings) to tissue damage caused by intense chemical (e.g., chili peppers), mechanical (e.g., pinching), or thermal stimulation; examples include postoperative pain, osteoarthritis-related pain, and mechanical low back pain. Neuropathic pain results from actual damage to the nervous system (e.g., diabetic peripheral neuropathy, poststroke pain and chemotherapy-induced neuropathy). Historically, opiates have been the primary class of medications used to treat chronic pain, but there are adverse consequences of chronic opiate therapy in people, especially those not receiving palliative care for terminal disease. Over half of those afflicted with chronic pain are dissatisfied with their current conventional treatments available to them. Survey reports suggest that patients are using cannabis to manage their pain with approximately 50 percent reduction in self-reported pain.[1] In fact, in a study of pain patients using opiates, those using cannabis were more likely to reduce their daily opiate subscription doses.[2]

THC and CBD and Pain

There is very good preclinical evidence that THC is effective in reducing pain in animal models of both acute and chronic pain. In such models, THC is effective when administered orally, systemically by injection, or directly into the brain or spinal cord, and most studies find that it reduces pain. Because of interactions between the endocannabinoid system and the opiate system, synergistic effects have been reported between THC and opiates in the regulation of pain. Low doses of THC have been found to significantly enhance morphine-induced analgesia when THC and either morphine or codeine are coadministered systemically by injection or orally in animal models.[3] Therefore, it is possible that combined treatments of THC and opiates may produce pain relief at doses that do not produce side effects on their own.

Pain relief is one of the classical "mouse tetrad symptoms" used to classify a drug as acting like THC; however, the psychoactive side effects of THC limit its usefulness in treating pain in patients. Preclinical studies have shown that CBD may also reduce chronic pain, although it appears to be ineffective in reducing acute pain in animal models. These effects were found to be mediated by the action of TRPV1 receptors (capsasin receptors—those acted upon by hot chili peppers) as the TRPV1-receptor antagonist (capsazepine) reversed CBD's attenuation of pain.[4] In addition, CBD has been shown to be effective in relieving neuropathic chemotherapy-induced pain in rats, and in diabetic mice, without the development of tolerance.[5] Because CBD is not intoxicating, it is a strong candidate for treatment of chronic pain. Human survey data suggest that chronic pain patients using CBD report pain relief and a reduction of opiate use; however, the findings from the few existing double-blind randomized control trials do not support these anecdotal claims.

Although few studies have evaluated the efficacy of CBD alone treatment in chronic pain patients, nabiximols (a sublingual spray of 2.5 mg THC/2.7 mg CBD per spray) has been evaluated in multiple clinical trials for pain management with generally favorable results. Because nabiximols is a combination of THC and CBD, it is most likely that THC is responsible for its pain-relieving effects, but CBD may modify some of the THC effects. Preclinical evidence suggests that CBD and THC may act synergistically to reduce pain.[6] Human RCTs revealed that nabiximols (with few side effects at low to moderate doses) was effective in treating pain and spasticity associated with multiple sclerosis and advanced cancer pain in opiate-resistant patients.[7] Nabiximols is approved for these uses in Canada and most European countries, but it has not been approved for this use by the US FDA.

Endocannabinoid System in the Regulation of Pain

Systemic administration of THC and synthetic CB_1-receptor agonists produce analgesia in animal models of acute and chronic pain. However, concerns about dependence, tolerance, and the cognitive side effects produced by global agonism of the CB_1 receptor and medicinal cannabis remain. There has been considerable preclinical research that has focused on potential manipulations of the natural endocannabinoid system in animal models. These manipulations do not produce the psychoactive side effects produced by global CB_1 activation. These potential treatments include CB_1 allosteric modulators, FAAH inhibitors, MAGL inhibitors, and CB_2 agonists.

CB_1 allosteric modulators. The natural endocannabinoids, 2-AG and AEA, inhibit pain transmission by acting on CB_1 receptors at central, spinal, and peripheral synapses.

Unlike THC, which acts globally on all CB_1 receptors, AEA and 2-AG are released on demand where and when they are needed and then are rapidly metabolized terminating their action. Their action is therefore more transient and selective under temporal and spatial regulation. Allosteric modulators of the CB_1 receptor bind to a distinct site apart from the primary orthosteric receptor site producing conformational changes in the receptor, which alters the potency of the endocannabinoid when it binds to the orthosteric site. Allosteric modulators have no physiological effect on their own in the absence of ligand (2-AG or AEA) binding. CB_1 positive allosteric modulators enhance the pain-relieving effects of endocannabinoids, but with limited side effects.[8]

FAAH and MAGL inhibitors. It is untenable to simply administer AEA or 2-AG systemically to reduce pain, because they are rapidly degraded by the enzymes FAAH and MAGL, respectively. One approach explored in preclinical research is treatment with inhibitors of the enzymes that degrade the endocannabinoids. Because of their "on-demand" production and release, endocannabinoids are specifically generated at bodily regions regulating pain to reduce the pain.

Animals that are genetically engineered to lack the enzyme FAAH (FAAH knockout mice) have abnormally high levels of AEA. These animals are less responsive to pain. As well, systemic administration of FAAH or MAGL inhibitors reduce acute and chronic pain by elevating AEA and 2-AG, respectively. Recently, to harbor the benefits of elevation of AEA and 2-AG, dual FAAH and MAGL inhibitors have been developed with the hope of enhancing the analgesic potential of these enzyme inhibitors. These manipulations have shown great promise in preclinical animal models but have not yet been shown to be effective in human RCTs.

Although the use of FAAH and MAGL inhibitors in preclinical studies has clearly supported the use of endocannabinoid-targeted compounds in clinical pain trials with humans, there has been only one published report of such a trial, and it was a failure. The Pfizer compound evaluated reduced FAAH activity by 96 percent and increased AEA levels substantially, but was not differentiated from placebo in reduction of osteoarthritis pain. However, it was well tolerated, with no evidence of cannabinoid-type adverse events. The targeted treatment in the Pfizer clinical trial, osteoarthritis pain, may differ qualitatively from the type of pain typically measured in many preclinical animal models that have demonstrated analgesic effects of FAAH inhibition. Most preclinical models measure reflex responses to a mechanical or thermal stimulus, whereas the predominant symptom in neuropathic pain evident in osteoarthritis is not evoked pain but spontaneous pain that is more difficult to model preclinically.[9] In view of the considerable preclinical evidence of the potential of FAAH inhibitors and MAGL inhibitors to reduce pain in a variety of models, there is a clear need to continue to evaluate the potential benefits of these treatments in other human models of chronic pain.

CB_2 agonists. Because THC is a mixed CB_1/CB_2-receptor agonist, another approach to the use of cannabinoids in treating chronic pain is the development of CB_2 agonists that only act on the nonpsychoactive CB_2 receptors that are primarily located in the peripheral nervous system. Unlike CB_1 agonists, CB_2 agonists are not psychoactive and thereby are devoid of central side effects. Activation of CB_2 receptors mediates the anti-inflammatory effects of endocannabinoids, which may be of particular importance in the development of chronic pain states. Treatments aimed at stimulating peripheral CB_2 receptors may reduce chronic arthritic pain.[10] The preclinical research results argue for the evaluation of CB_2 agonists in human RCTs, but none have yet been done.

How Effective Are Cannabinoids in Alleviating Chronic Pain in Humans?

Medical cannabis legalization has increased in recent years, and it is commonly used to treat chronic pain. Patient-reported data indicate that chronic pain management is one of the most common reasons for medical cannabis use. Furthermore, patients report that medical cannabis improves their pain-related outcomes, increasing the quality of their life and reducing their requirement for opioid analgesia. There are few alternatives for opioids in treating patients with chronic pain—only 40–60 percent of patients obtain even partial relief of their pain with current medication. Cannabinoids are a potential alternative treatment, even a low dose (1.29 percent THC) that does not produce psychoactive or cognitive side effects has been reported to be effective in reducing pain when assessed on a self-reported visual analogue scale.[11] There have been no clinical trials with humans of the efficacy of CBD alone for chronic pain.

It has been difficult to determine the effectiveness of cannabis to treat pain, because of the lack of high-quality double-blind, placebo-controlled clinical trials. Compared with the diversity of cannabinoids assessed preclinically, very few have been evaluated in human clinical trials. A better understanding of plant-derived cannabinoids, synthetic cannabinoids, and modulators of the endocannabinoid system in the treatment of pain is needed. Many of the available studies are limited by small sample size.

A recent report of the task force on cannabis and cannabinoid analgesia of the International Association for the Study of Pain (IASP), provided a comprehensive summary of the evidence from primary randomized controlled trials (RCTs) of cannabis (whole plant material), cannabinoids (extracts of cannabis constituents, such as THC and CBD, or synthetic THC or CBD) and cannabis-based medicines (extracts with regulatory approval as a therapeutic, such as nabiximols, dronabinol, Marinol, Epidiolex) in clinical acute and chronic pain management.[12] Randomized trials typically provide the least biased estimate of the treatment efficacy. The primary outcomes extracted from the study were as follows: (1) the proportion of people with at least 30 percent pain-intensity improvement defined as moderate, and (2) the proportion of people with at least 50 percent pain-intensity reduction defined as substantial. The study employed the Cochrane risk of bias tool that evaluates potential sources of bias including the randomization procedure for group assignment, possible selection bias, blinding of participants and personnel, attrition bias, reporting bias, and sample size bias (low risk of bias >200 participants/arm). Among a total of 36 RCT trials evaluated, only 2 included more than 200 participants/arm and were rated as low risk of bias for size. In many of the studies reviewed, pain conditions and doses of cannabinoids were not well defined and pharmacokinetic data were missing. The authors concluded that studies in this field have unclear or high risk of bias, and outcomes had ratings of low or very low-quality evidence with little confidence in the estimates of effect. In short, the evidence neither supports nor refutes claims of efficacy for cannabinoids, cannabis, or cannabis-based medicines in the management of pain.[13]

Patients report that medical cannabis improves their pain-related outcomes, increasing the quality of their life and reducing their requirement for opioid analgesia.

Even though patients report effectiveness, systematic reviews and meta-analyses generally report low to moderate levels of evidence to support the use of cannabis and cannabinoids for the treatment of chronic pain.[14] Some of the issues include limited availability of investigational products due to legal status, lack of standardization of cannabis products, lack of standardization of delivery of administration, and the nature of the defined outcomes of efficacy. Expert guidance on how to dose and administer medical cannabis safely and effectively is needed. Despite the low to moderate level of evidence of efficacy, patients are being treated with medical cannabis across the world.

Some studies have suggested that cannabis might preferentially target the affective aspects of pain, and therefore one of the beneficial mechanisms of cannabis use for pain disorders could be related to its ability to alter the reactive thoughts regarding negative feelings associated with pain. Cannabis might positively affect a patient's perception of pain and improve coping mechanisms. Recent clinical trials suggest that cannabinoid-mediated analgesia in humans could be attributed to a moderate reduction in affective response but not a reduced perception of the experimental pain.[15]

What about the effectiveness of use of cannabinoids in palliative care in patients with terminal cancer? One person in two or three who gets cancer will have moderate- or severe-intensity pain that tends to get worse as the cancer progresses. The World Health Organization recommends morphine-like medicines for moderate to severe pain from cancer, but one in six to ten people with cancer pain do not experience sufficient pain relief from morphine-like medicines. Some people with cancer pain have reported anecdotally that cannabis-based medicines are effective for them. A Cochrane review was recently reported of double-blind RCTs of medical cannabis, plant-derived, and synthetic cannabis-based medicines against placebo or any other active treatment for cancer pain in adults, with any treatment duration and at least ten participants per treatment arm.[16] The studies that met the Cochrane criteria for inclusion in the review found that cannabis-based medicines did not relieve cancer pain that did not respond to morphine-like medicines. The study concluded that trials with cannabis-based medicines in cancer need to be very much better designed than those conducted so far.

Yet medical cannabis appears to be useful for some patients for chronic pain. To assist clinicians who may have limited experience with prescribing medical cannabis, a recent report by twenty global experts recommending dosing and administration of medical cannabis to treat chronic pain has been recently published.[17] There was consensus that medical cannabis may be considered for patients experiencing neuropathic, inflammatory, nocioplastic, and mixed pain. Three treatment protocols (using oral administration—oils or gel capsules) were developed:

First, a routine protocol with CBD at a dose of 5 mg twice daily, increasing the dose by 10 mg every 2–3 days until the patient reaches their goals, or up to 40 mg/day. At a dose of 40 mg/day of CBD, clinicians may consider adding THC at a 2.5 mg dose and titrate by 2.5 mg every 2–7 days until a maximum dose of 40 mg/day of THC. Second, a conservative protocol where the clinician initiates a patient on a CBD dose of 5 mg once daily and titrates the CBD-predominant dose by 10 mg every 2–3 days until the patient reaches their goals or up to 40 mg/day. At a CBD dose of 40 mg/day, clinicians may consider adding THC at 1 mg/day and titrate by 1 mg every 7 days until a maximum daily dose of 40 mg/day of THC. Third, a rapid protocol where the clinician initiates the patient on a balanced THC:CBD variety at 2.5–5 mg of each cannabinoid once or twice daily and titrates by 2.5–5 mg of each cannabinoid every 2–3 days until the patient reaches their goal to a maximum THC dose of 40 mg/day.

CANNABINOIDS AND NEUROLOGICAL DISORDERS

Endocannabinoid signaling is implicated in multiple neurological disorders. It is involved in the regulation of cell, tissue, organ, and organism homeostasis as well as brain development, neurotransmitter release and synaptic plasticity, and cytokine release from microglia, all which are involved in a variety of neurological disorders. Substances that enhance or inhibit endocannabinoid activity can have therapeutic effects on neurological disorders in preclinical models, depending on disease characteristics and the roles of CB_1 and CB_2.[1]

Cannabinoids and Epilepsy

Epilepsy is a neurological disorder affecting approximately 1 percent of the world's population, characterized by recurrent, spontaneous seizures or convulsions due to disturbance of excitatory (glutamate)-inhibitory (GABA) equilibrium of neuronal activity in the brain. During a seizure, excitatory glutamate neurons fire excessively. Generalized seizures originate at a specific point within the brain but rapidly distribute across the brain to affect both hemispheres. Focal seizures are restricted to a specific region of the brain or a single hemisphere. Conventional antiepileptic drugs reduce the release of excitatory glutamate during a seizure by blocking sodium or calcium channels or enhancing GABA function, thereby preventing the spread of the seizure within the brain. These treatments are effective in approximately 50 percent of patients. All existing antiepileptic drugs are associated with numerous side effects (impairment of motor function, cognitive dysfunction, and emotional lability); therefore, there is a need for the development of better treatment options, including cannabinoids.

Endocannabinoid regulation of convulsions. Preclinical research has revealed that THC and other CB_1 receptor agonists can reduce seizures and CB_1 antagonists can exacerbate seizures in animal models. However, prolonged CB_1 activation may result in increased seizure activity—probably as a result of CB_1 receptor downregulation. The use of CB_1 agonists for treating epilepsy is therefore problematic.[2]

The endocannabinoid system acts as a homeostatic regulator of epileptic seizures. CB_1-expressing excitatory and inhibitory synapses regulate the dynamically changing normal and pathological oscillatory neural-network activity, with endocannabinoids controlling the activity of these networks. Endocannabinoids exert an overall protective effect against overexcitation from transient surges in activity in a brain that is not chronically epileptic (normal brain). In individuals with chronic ongoing epilepsy, this endocannabinoid-mediated negative feedback control system is functionally compromised and therefore is unable to prevent the generation of seizures. Temporal-lobe epilepsy, one of the most common forms in adults, involves changes in the expression of CB_1 receptors in the hippocampus. CB_1 receptors are present on both inhibitory GABA-containing terminals and excitatory glutamate-containing terminals in the hippocampus. The activation of CB_1 receptors on GABA terminals should be proconvulsive by reducing GABA release, and the activation of CB_1 receptors on glutamate terminals should be anticonvulsive by reducing glutamate release. Evidence from both patients and animal models indicates that as the disease progresses, upregulation of CB_1 receptors on GABAergic terminals (resulting in suppression of GABA) and downregulation of CB_1 receptors on glutamatergic terminals (resulting in enhancement of glutamate) mechanistically contribute to seizures.[3]

CBD as a Treatment for Seizures

Cannabis has been used as a treatment for seizures throughout history, particularly in ancient India and Assyria.[4] During the nineteenth century, several medical reports were published on the ameliorative effects of cannabis on several forms of convulsions, most notably as a treatment for Queen Victoria's seizures.[5] Of course, the use of whole plant extracts at the time did not reveal the phytocannabinoid responsible for this effect. Interestingly, it is not THC, but CBD, that has now been approved by the FDA as a potential treatment for seizures in epilepsy.

CBD has been shown to reduce seizures in preclinical models with rats and mice. This effect of CBD is not mediated by the action on CB_1 or CB_2 receptors but seems to be by modulation of intracellular calcium through interaction with TRPV1 receptors or GPR55 receptors.[6] The latter mechanism is particularly interesting. GPR55 receptors are located on excitatory (glutamatergic) axon terminals. When these receptors are activated, they *enhance* glutamate release when the neuron fires. Because CBD blocks GPR55 activation, it is ideal as an anticonvulsive agent. CBD would thereby dampen excess glutamate release from the hyperactive excitatory neurons during epileptic seizures. These preclinical studies set the stage for human clinical trial testing of CBD specifically for the treatment of epilepsy.

Raphael Mechoulam and Elisaldo Carlini conducted the first scientific human clinical trial on the efficacy of CBD to treat epilepsy in human patients.[7] In this small-scale (a total of 9 patients) placebo-controlled double-blind trial, patients received 200 mg of CBD daily ($n = 4$) or placebo ($n = 5$) for three months in addition to their normal anticonvulsion medication. Among the patients treated with CBD, two patients had no seizures for the entire 3-month period and one patient partially improved with no toxic effects reported for either group. These were very promising findings, but the sample size was very small. A subsequent study was conducted with fifteen patients who received 200–300 mg of CBD daily ($n = 8$) or placebo ($n = 7$) for up to 4.5 months in combination with their prescribed medications (which were no longer effective in controlling their symptoms).[8] CBD completely prevented seizures in 4 of the 8 patients, partly controlled seizures in 3 of the patients and was ineffective in 1 patient, with no severe side effects in any patients. Placebo was ineffective. Despite these initial promising results, which were consistent with the preclinical findings, it is only very recently that large-scale, double-blind, placebo-controlled clinical trials have been conducted, mostly in response to the anecdotal reports of the positive effects of CBD-rich cannabis on children with pediatric epilepsy. As indicated by Mechoulam et al., why did we have to wait for decades?[9]

Evidence indicating that CBD shows efficacy for treating epilepsy in humans was provided almost fifty years ago, yet until recently the published data on the use of CBD for the treatment of epilepsy came from fewer than seventy patients, with very few of them children. Few of these studies were rigorous, and few provided high-quality evidence.[10] However, recently stronger clinical evidence has been provided for the efficacy of CBD for the treatment of a highly resistant, rare form of epilepsy in children, Dravet syndrome. Beginning in the second year of life, children affected by Dravet syndrome develop an epileptic encephalopathy that results in cognitive, behavioral, and motor impairment. Research on the treatment of this condition has accelerated by widespread press coverage of anecdotal successes in 2016 (by Sanjay Gupta, chief medical correspondent for the television channel CNN).

The most well-known anecdotal report is that of Charlotte, a five-year-old girl in the United States who was diagnosed in 2013 with Dravet syndrome, with up to fifty generalized tonic-clonic seizures per day. Following three months of treatment with high-CBD strain cannabis extract (later marketed as "Charlotte's Web"), her seizures were reported to have reduced by more than 90 percent.

Two high-profile clinical trials resulted in the approval of CBD for childhood epilepsy. The first, which appeared in the *New England Journal of Medicine*, Devinsky et al. reported the results of a controlled trial of CBD treatment

Evidence indicated that CBD shows efficacy for treating epilepsy in humans was provided almost fifty years ago in children with Dravet syndrome.[11] In a double-blind, placebo-controlled trial, 20 children and young adults with Dravet syndrome were randomly assigned to receive either CBD oral solution (20 mg/kg/day) or placebo in addition to their standard antiepileptic treatment. CBD reduced the median frequency of convulsive seizures per month from 12.4 to 5.9 (with 5 percent of the sample becoming seizure free) as compared with a decrease from 14.9 to 14.1 with placebo. Adverse effects occurring more frequently in the CBD than the placebo group, included diarrhea, loss of appetite, and somnolence, with a withdrawal of 8 patients from the CBD group versus only 1 patient in the placebo group. Shortly thereafter, the results of a second clinical trial appeared in the *Lancet*, which evaluated the efficacy and safety of CBD in treating a group of children with Lennox-Gastaut syndrome, via an epileptic encephalography that produces various types of treatment-resistant seizures, including drop seizures.[12] This multisite study with 24 clinical sites in the United States, the Netherlands, and Poland included a total of 171 patients that were randomized to receive CBD (200 mg/kg oral solution, $n = 86$) or matched placebo ($n = 85$) along with their normal antiepileptic drug regime. Treatments were given daily for 14 weeks with 2 weeks of dose escalation and 12 weeks of maintenance at the 200 mg/kg dose. At the end of the treatment, a 10-day dose taper was implemented. The results of this trial revealed that the

CBD group decreased drop seizure frequency by a median of 43.9 percent compared with 21.8 percent in the placebo group. Other types of seizures were also reduced in the CBD group by a median of 49.4 percent compared with 22.9 percent in the placebo group. Treatment-related adverse effects (diarrhea, somnolence, and reduced appetite) were more frequent in the CBD group than in the placebo group. These successful trials resulted in the approval of a pure oral CBD product (Epidiolex) for these conditions by the FDA in the United States in 2018.

Epidiolex is undergoing clinical trials in adults for a broader range of epilepsy syndromes with very positive results; however, some recent findings suggest that CBD interferes with the metabolism of another epileptic drug, clobazam, thereby elevating the plasma concentration of clobazam's major active metabolite.[13] This led to the suggestion that CBD may reduce seizures indirectly by enhancing the efficacy of other coadministered antiepileptic treatments. However, subsequent clinical trials verified that such drug-drug interactions cannot account for the antiseizure effects of CBD.[14] A recent comprehensive review describes the interactions between CBD and other medications, illicit substances, and alcohol.[15] Future clinical trials will need to take these drug interactions into account.

The strongest clinical evidence for the efficacy of CBD as an adjunct therapeutic agent is in the treatment of rare childhood epilepsy. Epidiolex is an FDA-approved medical

Successful human clinical trials resulted in FDA approval of a pure oral CBD (Epidiolex) for treatment of childhood epilepsy syndromes.

treatment for these disorders. Further clinical trials are essential to determine whether CBD is also an effective treatment for epilepsy in adulthood and perhaps also as a treatment for nonepileptic seizures, given the range of seizure types associated with Dravet's syndrome.

Cannabinoids and Neurodegenerative Disorders

A neurodegenerative disorder involves progressive loss of neurons, often resulting from oxidative stress and neural inflammation. The consequences of neurodegeneration (ranging from movement disorder to dementia) can vary widely depending upon the specific region of the brain affected. Neurodegenerative disorders include such disease states as multiple sclerosis (MS), Alzheimer's disease (AD), Huntington's disease (HD), and Parkinson's disease (PD). These conditions are characterized by progressive deterioration of neurons in the central nervous system. They result in debilitating symptoms and a decline in cognitive and motor functions. Effective treatment for these conditions remains elusive despite extensive research. Current treatments mainly target symptoms, and there are no current pharmacotherapies to prevent neurodegeneration or induce neural repairing.[16]

Preclinical studies have shown that cannabinoids may be useful treatments for neurodegenerative disorders because they protect neurons from damage and promote neuronal survival. Cannabinoids protect not only neurons but also some glial cell subpopulations from various insults, including those involved in neurodegenerative disorders such as MS. Preclinical studies suggest that they may delay or arrest disease progression. By their action on CB_1 and CB_2 receptors, cannabinoids reduce excitatory glutamate release, reduce intracellular calcium, produce antioxidant effects, and produce anti-inflammatory effects—all neuroprotective effects. Thus, cannabinoids have an advantage of combining several neuroprotective mechanisms that may be particularly important for neurodegenerative disorders in which neuronal damage is the consequence of different types of cytotoxic events.[17]

Several preclinical and clinical studies have described an unbalance in the endocannabinoid system in animal models and patients diagnosed with AD, PD, HD, and MS.[18] Also, studies focusing on animal models of neurodegenerative diseases showed that modulation of the endocannabinoid system is a valid alternative to improve animal's conditions.[19] The most important neuroprotective property of CB_1 agonists is normalization of glutamate homeostasis. Excessive glutamate activity results in intracellular accumulation of calcium, which activates numerous destructive pathways that lead to cell swelling and death. Activation of CB_1 receptors opposes glutamatergic cytotoxic events.

Endocannabinoids oppose the effects of stimuli that damage the brain. Activation of the endocannabinoid system has been observed in some neurodegenerative conditions, including brain trauma in neonatal and adult rats, experimental Parkinsonism in rats, and kainate-induced excitotoxicity in mice.[20] The upregulation (compensatory increase) of CB_1 receptors has been observed after experimental stroke produced by excitotoxic stimuli in neonatal rats and in the postmortem basal ganglia of Parkinson's disease patients.[21]

CB_2 receptors display a marked upregulation in all neurodegenerative disorders, including AD, PD, HD, and MS. In the normal healthy brain, CB_2 receptors are located primarily on glial cells, including astrocytes and oligodendrocytes. They are absent from nonactivated microglial cells. These CB_2 receptors upregulate in response to inflammatory, excitotoxic, infectious, traumatic, or oxidant insults occurring in most neurodegenerative disorders, and this upregulation of CB_2 receptors is extremely intense in reactive microglial cells recruited to lesioned sites. Targeting CB_2 receptors, alone or in concert with CB_1 receptors, may improve neuronal homeostasis in neurodegenerative disorders.[22] However, there have been no human clinical trials with CB_2 receptor agonists as treatments for neurodegenerative diseases.

Most of the evidence to date for the potential of cannabinoids to protect against the neurogenerative disorders of AD, PD, and HD are based on preclinical animal studies. Human clinical studies evaluating the effects of cannabinoids on these neurodegenerative disorders are limited. A recent systematic review evaluating the use of phytocannabinoids in AD concluded that these compounds were effective in alleviating some secondary symptoms of this disorder, but not in aspects involving memory and cognition.[23] Despite the positive preclinical findings regarding the effects of CBD in reducing neuroinflammation of this disorder, there are currently no clinical trials to evaluate the potential of CBD in the control of AD symptoms in patients. Among HD patients, the results of clinical studies with nabilone (synthetic THC) have described some improvement in motor symptoms.[24] The first clinical trial that actually monitored the progression of HD (rather than the affect on symptom relief) was conducted with nabiximols (Sativex, 2.7 mg THC/2.5 mg CBD/spray). Unfortunately, there was no evidence of slower disease progression, but the clinical trial demonstrated that the medicine was safe and was well tolerated by HD patients.[25] Finally, in a recent systematic review of the literature, five randomized controlled studies and eighteen nonrandomized studies investigated cannabis treatment in PD patients.[26] No compelling evidence was found to recommend the use of cannabis in PD, although it appeared to reduce some of the symptoms including anxiety and pain. In particular, CBD appears to reduce anxiety and tremor in PD patients in a simulated public speaking environment.[27] Given the paucity of well-designed randomized studies, there is a need for further investigation of cannabinoid treatments for AD, PD, and HD.

The most promising evidence for the use of cannabinoids in treatment of neurodegenerative disorders is found in the use of nabiximols in treating the symptoms of spasticity and chronic pain in MS patients. In fact, recently, nabiximols has been approved for prescription for MS in Canada and several European countries for this purpose.

Cannabinoids and Multiple Sclerosis

Multiple sclerosis (MS) is an inflammatory disease of the brain and the spinal cord, resulting in damage of myelin and axons that affects 2–3 million people in the world. It is three times as common in females than in males. Early in the course of the disease, inflammation is transient and remyelination occurs; therefore, patients usually recover from symptoms of neurological dysfunction. Over time, axonal loss produces progressive disability in MS, due to widespread microglial activation associated with chronic neurodegeneration. The levels of glutamate in the cerebrospinal fluid of MS patients undergoing an inflammatory episode suggests a loss of homeostatic control of neurotransmission resulting in excess activity of excitatory circuits due to a loss of inhibitory circuits under the control of endocannabinoids.[28]

Cannabinoids and MS Symptoms

Among the symptoms of MS that patients claim cannabis alleviates are bladder incontinence, tremor, limb spasticity, and pain. Spasticity, one of the most commonly reported symptoms, may affect approximately 50 percent of patients, for which current therapies (e.g., benzodiazepines, baclofen) are relatively ineffective and have adverse side effects. The pathophysiology of spasticity is not well understood, but it may reflect a loss of inhibitory circuitry in the spinal cord that can result in excessive contractions of the muscles, sometimes even when the patient is at rest. In mouse models of multiple sclerosis, nabiximols (THC:CBD) alleviated hind-limb spasticity. As well, administration of FAAH inhibitors (which elevate AEA) also reduces the level of spasticity in mice. Therefore, the preclinical data provide clear evidence that cannabinoids have therapeutic potential for MS.[29]

There is relatively high-quality evidence in the human clinical literature for the use of cannabinoids to treat chronic pain and spasticity in MS.[30] Recently, nabiximols has been approved for prescription for MS in Canada and in several other Western countries, but not in the United States. Nabiximols is used in Canada, the United Kingdom,

The most promising
evidence for the use
of cannabinoids
in treatment of
neurodegenerative
disorders is found in the
use of nabiximols in
treating the symptoms
of spasticity and chronic
pain in MS patients.

and Spain to treat spasticity associated with MS, and in Canada it also is used to treat neuropathic pain in MS and as an add-on treatment to strong opioid therapy in patients with advanced cancer. There is no evidence of tolerance or withdrawal from termination of nabiximols.

Are Cannabinoids Neuroprotective in MS?

There is increasing evidence that elevated levels of glutamate are seen not only in MS patients but also in animals exhibiting MS spasticity. A reduction of the effects of elevated glutamate shows disease amelioration in experimental studies and clinical studies. Endocannabinoids may have neuroprotective properties in neuroinflammatory disease by downregulating the release of glutamate. In CB_1 knockout mice, neuroinflammation resulted in accelerated accumulation of neurological deficits, and administration of CB_1 agonists in wild-type mice can inhibit neurodegeneration in animal models of MS. These findings suggest that cannabinoid therapy may have potential to slow neurodegeneration in MS patients and may be considered as an adjunct therapy to current disease-modifying therapies.[31]

Based on promising preclinical findings, a RCT was conducted to determine whether cannabinoid therapy would slow the neurodegeneration that causes the progression of MS in humans.[32] A total of 493 MS patients who had disrupted walking were randomly assigned to receive THC capsules or placebo capsules over a period of three years. Despite the abundant preclinical experimental evidence suggesting that THC has a neuroprotective role in neurological diseases, the study found no evidence that THC affected the progression of MS. There was some evidence, however, that in participants at the lower end of the disability scale THC had statistically significantly beneficial effects on the neurological assessments by physicians relative to participants given placebo. Because this represented a small percentage of the participants, further studies are necessary targeting patients at the lower end of the disability spectrum to determine whether THC can protect against MS progression.[33]

Cannabis and medications derived from it have been confirmed to alleviate MS-related spasticity in both experimental and clinical settings, and nabiximols has been added to the list of medications in various countries. An issue that has not been resolved in many countries is the economic cost of the drug to a patient suffering from MS. As with all cannabis medications, efficacy will have to be balanced with the well-known side effects of cannabis caused by global stimulation of cannabinoid receptors in the brain rather than stimulation only in the regions needed to subside the symptoms—effects that many patients find undesirable.

CONCLUSIONS AND WHAT'S NEXT

Aside from alcohol, cannabis is the most widely used recreational drug in the Western world. Cannabis is consumed by approximately 83 million people, which is nearly 3 percent of the world population.[1] Driven by legalization and medical use in many countries and states in the United States, there has been a change in the risk perception of cannabis. As well, compelling research evidence and subsequent FDA approval of CBD for severe childhood epilepsy have confirmed the therapeutic potential of CBD. It is becoming increasingly clear that our body has a complex endocannabinoid system upon which phytocannabinoids act. Once these endocannabinoids were discovered over thirty years ago, elements of the endocannabinoid system have been the target of new drugs aimed at treating several human diseases.[2]

Risks of Recreational Cannabis

Although the risk perception of recreational cannabis has softened considering its increased accessibility in recent years, certain health concerns have remained paramount to some professionals. In particular, the putative link between adolescent cannabis use and subsequent psychotic disorders. The link may be correlational rather than causal—as there is a suggestion that adolescents prone to develop psychotic disorders may also be those prone to use cannabis, or a third variable, for instance, proneness to risk-taking may underly both tendencies. However, clearly the evidence suggests that individuals with a family history of psychosis or who have experienced acute psychosis under the influence of cannabis would be advised to avoid using cannabis in adolescence, in particular. Cannabis use has also been suggested to interfere with adolescent cortical brain development during a critical period. It is essential that high-quality trials be continued to ascertain the validity of these potential risks of cannabis use to the developing adolescent brain.

For many years, it has been debated whether cannabis was addicting like other recreational drugs. It is now clear that "cannabis use disorder" is a potential health concern for individuals chronically administering cannabis. Clearly, individuals diagnosed with "cannabis use disorder" experience withdrawal symptoms upon termination of cannabis (including irritability, depression, anxiety, and sleepless-ness), although they are less severe than those from some other recreational drugs such as opiates. Cannabis use disorder is accompanied by a persistent desire or unsuccessful effort to cut down or stop using cannabis. A great deal of time is devoted to the pursuit of cannabis with craving upon termination of use. Approximately 25 percent of individuals presenting for help in addiction centers are there for cannabis use disorder.[3]

The increasing misconception among the general public that cannabis is "harmless" abates several other potential harms of cannabis, which include prenatal effects on the fetus, potential cardiovascular and respiratory effects on the user, and dangers of driving under the influence. Results from the longitudinal NIH Adolescent Brain and Cognitive Development (ABCD) Study revealed that women who continued to use cannabis after learning they were pregnant (compared with those who did not use cannabis or those who stopped using cannabis after learning they were pregnant) had babies with lower birth weight, lower total intracranial white matter volumes, and greater later cognitive deficits.[4] Clearly, cannabis should not be used by pregnant women. Cannabis can also be harmful to users themselves. Just as tobacco smoke is harmful to the lungs, so, too, is cannabis smoke. Cannabis also produces tachychardia, speeding up heart rate, and can produce cardiovascular problems in rare individuals. Cannabis use

while driving produces cognitive deficits as revealed by driver-simulation programs.

Benefits of Medical Cannabis

Medical cannabinoid products (plant-derived and synthetic derivatives) differ in their pharmacology. Synthetic THC (dronabinol and nabilone) acts as partial agonist of the CB_1 and CB_2 receptors that is approved by the FDA for HIV/AIDS-induced loss of appetite and chemotherapy-induced nausea and vomiting. CBD (Epidiolex) acts at several receptors and has been approved in the United States and Europe for treatment of seizures in childhood epilepsy. Nabiximols, a cannabis-based extract containing equal quantities of THC and CBD (Sativex), was approved in 2010 in the United Kingdom for symptoms associated with MS, and exported to more than 28 countries from Asia, Africa, the Middle East, Europe (Spain, the Czech Republic, Germany, Denmark, Sweden, Italy, Austria, France, and Poland) and Canada (but not the United States). It is also known that cannabis contains over 125 phytocannabinoids, most of which have not been pharmacologically characterized or even isolated for that matter. The concentration of THC and CBD in various medical cannabis products varies in clinical trials conducted for a variety of conditions. Therefore, systematic reviews of the effectiveness of various cannabinoid treatments for therapeutic efficacy may be misleading.

Numerous "cannabinoid products" that are not approved by regulatory agencies are being tested for their potential therapeutic value. It is difficult to discern the composition and concentration of active agents in these herbal preparations, which are variously described as cannabis, cannabis oil, smoked cannabis (cigarettes), inhaled cannabis, vaporized cannabis, cannabis extract, or "CBD-rich"/"THC-rich" marijuana or extracts. The challenges of determining the active synergistic agents, appropriate dosing schedule, specificity of therapeutic use, and safety profile remain to be overcome when herbals are used as medicinal products.

A recent pharmacology-based systematic review of specific cannabinoids for all relevant medical conditions revealed that the confidence on the effect estimate strongly differs for the type of cannabinoid evaluated.[5] Among the various cannabinoids reviewed, the highest quality of evidence was found for CBD in the treatment of epilepsy for Dravet syndrome and Lennox-Gastaut syndrome. Second, the evidence for the effects of cannabinoids for chronic pain revealed that only dronabinol and nabiximols had moderate evidence for effectiveness. Third, dronabinol for appetite stimulation in AIDS patients revealed moderate evidence for effectiveness. Fourth, CBD for Parkinson symptoms had a moderate effect. Fifth, nabiximols appeared to have a moderate effect on sleep quality. Finally, dronabinol and nabilone reduced nausea and vomiting with a low level of evidence for effectiveness, but CBD alone has not been tested and two studies suggested that nabiximols were effective for nausea and vomiting.[6] Another recent meta-analysis revealed that THC (dronabinol and nabilone) has no effect on appetite, whereas CBD decreases appetite with moderate evidence.[7]

An extensive systematic review of seventy-nine randomized controlled trials using the Cochrane Collaboration's tool for assessing risk of bias evaluated the benefits and adverse effects associated with medical cannabis across a broad range of conditions (the majority being nausea and vomiting in chemotherapy or chronic pain and spasticity due to MS).[8] With a moderate risk of bias, the evidence was strongest that smoked cannabis and oromucosal THC:CBD mixtures may be beneficial for the treatment of chronic neuropathic or cancer pain and that nabiximols, nabilone, THC/CBD capsules, and dronabinol may be beneficial for the treatment of spasticity due to multiple sclerosis. Although the cannabinoids also were effective for chemotherapy-induced nausea and vomiting, the risk of bias among these studies was high. Interestingly there was no clear evidence for a difference in effectiveness or in adverse effects based on the type of cannabinoid or the mode of administration.

The Future of Medical Cannabinoids—Manipulation of the Endocannabinoid System

As we see throughout this book, agonists of the CB_1 receptor are critical for our bodies to maintain a homeostatic balance in several physiological systems—this is how AEA appears to regulate emotional stress in our brain. However, agonists of the CB_1 receptor as medicines have the often-unwanted side effects of psychoactivity sought out by recreational cannabis users. Developing treatments that act at CB_1 receptors without these psychoactive side effects is a challenge in drug development. One approach has been the development of CB_1 agonists that are peripherally restricted with limited access to brain CB_1 receptors.[9] Another effort to address selectivity is found in newly developed allosteric modulators of the CB_1 receptor. Allosteric ligands modify the conformation of the receptor protein, which allows for modulating the affinity of the orthosteric ligands. Allosteric ligands can either augment (positive allosteric modulators) or diminish (negative allosteric modulators) the effect of endogenous ligands. Positive allosteric modulation is likely to be increasingly important in drug development. For instance, ZCZ011 increased the potency and reduced tolerance development in the antinociceptive activity of CB_1 agonists.[10]

The therapeutic value of CB_1 receptor antagonists/inverse agonists, such as rimonabant, in metabolic disorders and addiction highlighted the efficacy of blocking CB_1 receptors for several conditions with unmet medical needs (including obesity and addiction); however, psychiatric side effects prevented their application to medical treatment. Additional approaches are being explored to retain their efficacy and circumvent the unwanted neuropsychiatric side effects of depression and anxiety. Among these, CB_1 receptor antagonist/inverse agonists that cannot enter the CNS and CB_1 receptor neutral antagonists without inverse agonist properties have shown promising results in preclinical models of obesity and addiction.[11]

The therapeutic potential of CB_2 receptor agonists is likely to play an important role in future cannabinoid treatments.[12] Multiple CB_2 selective agonists and antagonists have been synthesized and evaluated in preclinical animal models. As well, a few CB_2 allosteric ligands have also been discovered. The CB_2 receptor is primarily expressed in immune cells and is highly inducible in microglia upon neuroinflammation. Future CB_2 receptor ligands have demonstrated huge therapeutic potential in a large variety of disease models and pain.[13] Generally, the reported effects are a consequence of CB_2 mediated immunosuppressive and anti-inflammatory effects leading to a dampening of tissue injury. Several current clinical trials are ongoing with CB_2 receptor agonists. Overall, more than 20 new CB_2 receptor agonists have been investigated in humans for a wide range of indications.[14] Ten additional

The main interest in this field is to develop therapeutic alternatives to the use of CB_1 receptor agonists, able to prevent or minimize serious psychotropic side effects due to direct receptor activation.

new CB_2 agonists were investigated in phase 1 and 2 clinical trials for different pain indication (neuropathic, dental, and osteoarthritis of the knee), posttherapeutic neuralgia, stroke, traumatic brain injury, with no CB_2 adverse effects being reported.[15]

There is also considerable therapeutic potential of inhibition of the primary metabolic enzyme of AEA and other fatty acids, FAAH. The FAAH inhibitor, URB597, which unlike THC is not rewarding to nonhuman primates, has shown to be beneficial for anxiety disorders, substance use disorders, and pain in preclinical animal models.[16] A few recent phase 2 clinical trials have shown some success using FAAH inhibitors as treatments for anxiety and cannabis use disorder. In a multicenter, placebo-controlled phase 2 trial, patients with social anxiety disorder that were treated with a FAAH inhibitor showed evidence of reduced anxiety, but the dosage was too low to fully inhibit FAAH.[17] Additional clinical testing in anxiety is clearly warranted. As well, a phase 2 clinical trial also demonstrated that a FAAH inhibitor was effective in reducing cannabis use and alleviating withdrawal symptoms in men with cannabis use disorder.[18]

As well, inhibition of the primary metabolic enzyme of 2-AG, MAGL, has shown preclinical promise in treating cancer, PD, AD, MS, inflammatory and neuropathic pain, and anxiety/depression. However, unlike FAAH inhibitors, MAGL inhibitors cause desensitization of CB_1

and behavioral tolerance to CB_1 agonists, suggesting the possibility of psychoactive effects in humans, which limits their usefulness. New compounds are under development to circumvent these possible limitations.

The main interest in this field is to develop therapeutic alternatives to the use of CB_1 receptor agonists, able to prevent or minimize serious psychotropic side effects due to direct receptor activation. The possibility to increase eCB tone by reducing degradation of AEA or 2-AG and to apply new multitarget strategies that include additional receptors and enzymes have boosted FAAH and MAGL inhibition studies to generate therapeutics against peripheral and CNS-related pathologies. These new technologies are likely to produce new medicines that take advantage of the new scientific advances in our understanding of the endocannabinoid system that appears to play such an important role in regulating the functioning of our body.

ACKNOWLEDGMENTS

I would like to thank Dr. Erin Rock, Neuroscience Research Manager, University of Guelph, for her assistance in formatting the References and the Notes for the final document, and Jolanta Komornicka for preparing the index, as well as Anne-Marie Bono, Caroline Helms, and Elizabeth Agresta at the MIT Press for feedback throughout the preparation of the manuscript.

ACKNOWLEDGMENTS

Δ^9-tetrahydrocannabidiol (THC)
The principal psychoactive compound found in cannabis.

Δ^9-tetrahydrocannabidiol acid (THCA)
The biosynthetic precursor of Δ^9-tetrahydrocannabidiol (THC) in the cannabis plant.

2-arachidonoyl acid (2-AG)
An endocannabinoid that acts on Cannabinoid 1 (CB_1) and Cannabinoid 2 (CB_2) receptors.

Affinity
The strength of binding of a ligand to a receptor.

Agonist
A substance that binds to a receptor and mimics the actions of the endogenous substance that normally binds to the receptor.

Allosteric modulator
Allosteric modulators are defined as ligands that bind to a site on the receptor that is topographically distinct from the orthosteric binding site. These ligands modify receptor conformation to cause a change in the binding or functional properties of orthosteric ligands.

Alzheimer's disease (AD)
A neurodegenerative disease that produces dementia, most commonly expressed as difficulty in remembering recent events.

Anorexia nervosa
An eating disorder characterized by food restriction, body image disturbances, fear of gaining weight, and a desire to be thin.

Antagonist
A type of receptor ligand that blocks or dampens a biological effect by binding to the receptor.

Anxiogenic
Anxiety-producing drug/treatment.

Anxiolytic
Anxiety-reducing drug/treatment.

Anandamide (AEA)
An endocannabinoid that acts on Cannabinoid 1 (CB_1) and Cannabinoid 2 (CB_2) receptors.

Anterograde transmission
Movement of the neurotransmitter from the presynaptic neuron to the post-synaptic neuron.

Bioavailability
The fraction of an administered dose of an unchanged drug that reaches systemic circulation.

Bulimia nervosa
A type of eating disorder in which people eat large amounts of food at one time (binge) and then try to get rid of it (purge).

Cachexia
Chronic wasting disorder associated with loss of adipose tissue and lean body mass seen in patients with cancer, acquired immunodeficiency disorder (AIDS), and anorexia nervosa.

Cannabidiol (CBD)
The primary nonintoxicating compound found in cannabis.

Cannabidiolic acid (CBDA)
The precursor to CBD found in the cannabis plant before it is heated.

Cannabidivarin (CBDV)
A cannabidiol-like compound found in cannabis.

Cannabigerol (CBG)
The "mother of all cannabinoids" from which all other cannabinoids in the plant (including THC and CBD) are derived.

Cannabigerol acid (CBGA)
The precursor to CBG found in the cannabis plant.

Cannabinoid 1 (CB$_1$) receptor
A receptor upon which endocannabinoids (anandamide and 2-AG) as well as THC acts. Action at the CB$_1$ receptor in the brain produces psychoactive effects of cannabinoids.

Cannabinoid 2 (CB$_2$) receptor
A receptor upon which endocannabinoids (anandamide and 2-AG) as well as THC acts. Action at the CB$_2$ receptor does not produce psychoactive effects of cannabinoids. CB$_2$ receptors are primarily found in the immune system of the body.

Cannabis hyperemesis disorder
Cannabis hyperemesis disorder is characterized by severe cycles of nausea, vomiting, and abdominal pain in response to chronic use of high-dose THC.

Cannabis use disorder
Cannabis use disorder, also known as cannabis addiction, is a psychiatric disorder defined in the fifth revision of the *Diagnostic and Statistical Manual of Mental Disorders* as the continued use of cannabis despite clinically significant impairment.

Catechol-O-methyl-transferase (COMT)
An enzyme that metabolizes dopamine in the prefrontal cortex. Val/Val genotype is a potential genetic marker to predict likelihood of developing schizophrenia in response to cannabis use in adolescence.

Catalepsy
The period of immobile posture observed in mice, where they place their forepaws on a horizontal bar and hind paws on the ground.

Central nervous system (CNS)
The nervous system consisting of the brain and the spinal cord.

Cholecystokinin (CCK)
A satiety hormone in the gut.

Chromatography
The separation of a mixture by passing it in solution or suspension or as a vapor (as in gas chromatography) through a medium in which the components move at different rates.

Cochrane review
Cochrane reviews base their findings on the results of studies that meet certain quality criteria, because the most reliable studies will provide the best evidence for making decisions about health care.

Cognition
The mental action or process of acquiring knowledge and understanding through thought, experience, and the senses.

Cytochrome P450 liver enzymes
Membrane-bound enzymes that catalyze the monooxygenation of a diverse array of foreign (drug) and endogenous compounds. They are generally localized in the endoplasmic reticulum of the liver, lung, and small intestine.

Decarboxylation
Decarboxylation is the process of applying heat to cannabis to convert THCA and CBDA to THC and CBD, respectively.

Diacylglycerol (DAG)
Precursor to 2-AG found in the postsynaptic neuron.

Dravet syndrome
A highly resistant, rare form of epilepsy in children. Beginning in the second year of life, children afflicted by Dravet syndrome develop an epileptic encephalopathy that results in cognitive, behavioral, and motor impairment.

Drug-taking
Responding to access for the drug in preclinical models.

Drug-seeking
Responding to access for a cue previously paired with a drug in preclinical models as a model of drug-craving.

Dysphoria
A sense of unease with life.

Endocannabnoid system
Endogenous system consisting of the endocannabinoids, anandamide, and 2-AG, which act on CB_1 and CB_2 receptors and are degraded by fatty acid amide hydrolase (FAAH) and monoacylglycerol (MAGL) enzymes.

Endorphin
Endogenous peptide chemicals in the brain and nervous system, which act on the body's opiate receptors.

Efficacy
The power of an agonist to produce a pharmacological response.

Entourage effect
The action of the whole cannabis plant consisting of the combined effect of numerous cannabinoid compounds.

Epidiolex
A formulation of pure plant-derived CBD that has been approved by the FDA for treatment of childhood epilepsy.

Extinction
Extinction refers to the gradual weakening of a conditioned response that results in the behavior decreasing or disappearing. In other words, the conditioned behavior eventually stops.

Fatty acid amide hydrolase (FAAH)
Fatty acid amide hydrolase (FAAH) is an integral membrane enzyme that hydrolyzes the endocannabinoid anandamide and related amidated signaling lipids.

Fatty acid binding (FAB) protein
A group of proteins that coordinate lipid transport and signaling in cells.

First-pass metabolism
When a drug is orally consumed, it must first pass through the gut and the liver where enzymes act to metabolize it before entering the systemic blood stream.

GABA (Gamma-aminobutyric acid)
A neurotransmitter that acts at inhibitory GABA receptors.

Genome-wide association studies (GWAS)
An approach that involves rapidly scanning markers across complete sets of DNA or genomes of many people to find genetic variations associated with a particular disease.

Ghrelin
A hormone that stimulates feeding.

Glial cells
Glia, also called glial cells (gliocytes) or neuroglia, are nonneuronal cells in the central nervous system (brain and spinal cord) and the peripheral nervous. They hold nerve cells in place and help them work the way they should.

Glutamate
A neurotransmitter that acts at excitatory glutamate receptors.

G-protein receptor 55 (GPR55)
A G-protein coupled receptor with the endogenous ligand lysophophatidylin-sitol (LPI); CBD acts as an antagonist at this receptor.

Hashish
A cannabis concentrate product composed of compressed or purified preparations of stalked resin glands, called trichomes, from the plant.

Half-life
Time to excrete half of a dose.

Homeostatis
The steady state of internal equilibrium of physical and chemical conditions maintained by living organisms.

Huntington's disease (HD)
An inherited neurodegenerative disease resulting in a lack of coordination and unsteady gait, accompanied by cognitive decline.

Hypothalamus-pituitary-adrenal (HPA) axis
A complex set of direct neural and hormonal influences as well as feedback interactions among the hypothalamus, pituitary gland, and adrenal glands that control reactions to stress.

Inverse agonist
A drug that binds to the same receptor as an agonist but induces a pharmacological response opposite to that of the agonist.

Lennox-Gastaut syndrome
Epileptic encephalography that produces various types of treatment-resistant seizures, including drop seizures.

Leptin
A hormone predominantly made by adipose tissue cells with a primary role to regulate energy balance, appetite, and satiety.

Ligand
An endogenous molecule that binds with a specific receptor.

Long-term potentiation (LTP)
A persistent strengthening of synapses based on recent patterns of activity.

Long-term depression (LTD)
An activity-dependent reduction in synaptic strength.

Lysophosphitidylinositol (LPI)
An endogenous ligand for GPR55 receptors.

Memory consolidation
Memory consolidation refers to the process by which a temporary, labile memory is transformed into a more stable, long-lasting form.

Mesolimbic dopamine system
A central nervous system circuit in which dopaminergic inputs from the ventral tegmental area (VTA) innervate brain regions involved in executive, affective, and motivational functions, including the prefrontal cortex (PFC), amygdala, and nucleus accumbens (NAc).

Microglia
Resident immune cells of the brain that patrol for pathogens and damage.

Monoacylglycerol lipase (MAGL)
An enzyme that deactivates 2-arachidonoyl glycerol (2-AG).

Mouse tetrad assay
A series of behavioral paradigms used to show THC-like effects in mice. The four behavioral components of the tetrad are spontaneous activity, catalepsy, hypothermia, and analgesia.

Multiple sclerosis (MS)
An inflammatory disease of the brain and spinal cord resulting in damage of myelin and axons.

Nabiximols (trade name Sativex®-GW Pharmaceuticals)
A cannabis extract available as a sublingual spray consisting of 2.5 mg THC/2.5 mg CBD per spray that is approved for symptoms of multiple sclerosis in several countries.

N-arachidononlyl phosphatidylethanolamide (NAPE)
Endogenous precursor of anandamide (AEA).

Neurodegenerative disorder
A progressive loss of neurons.

Neuropathic pain
Pain caused by a lesion or disease of the nervous system; may be produced by normally nonpainful stimuli (allodynia).

Neurotransmitter
A signaling molecule secreted by a neuron to act on a receptor on another cell across the synapse.

N-methyl-D-aspartate (NMDA) glutamate receptor
A glutamate receptor found in neurons critical for synaptic plasticity in long-term potentiation (LTP).

Nociceptive pain
Pain caused by stimulation of sensory nerve fibers that respond to stimuli approaching or exceeding harmful intensity, including thermal, mechanical, and chemical.

Orthosteric binding site
The site on the receptor to which the endogenous ligand binds (the main receptor site).

Palliative care
An interdisciplinary medical caregiving approach aimed at optimizing quality of life and mitigating suffering for people with terminal illness.

Parkinson's disease (PD)
A neurodegenerative disease of the central nervous system that affects motor systems of the body and cognition in advanced stages. Pathophysiology is characterized by progressively expanding nerve cell death originating in the substantia nigra, a midbrain region that supplies dopamine to the basal ganglia, a system involved in voluntary motor control.

Peak plasma concentration (C_{max})
The highest concentration of a drug in the bloodstream after administration that is critical for evaluating its safety and effectiveness.

Peripheral nervous system (PNS)
The nervous system outside the brain and the spinal cord.

Phytocannabinoid
Specific molecules that come from the cannabis plant.

Preclinical animal model
Animal models of human disease.

Prodromal symptoms
A medical term for early signs or symptoms of an illness or health problem that appear before the major signs or symptoms start.

Prospective study
A study in which individuals are followed over time and data about them are collected as their characteristics or circumstances change.

Psychoactive
A drug or other substance that affects how the brain works and causes changes in mood awareness, thoughts, feelings, or behaviors.

Randomized control trial (RCT)
A study in which the participants are divided by chance into separate groups that compare different treatments or other interventions. Using chance to divide people into groups means that the groups will be similar and that the effects of the treatments they receive can be compared more fairly.

Receptor
A molecule on the surface or inside a cell that binds to a specific substance and causes a specific effect in the cell.

Receptor desensitization or downregulation
The process by which the concentration and affinity of receptors are decreased. Increased exposure to an agonist can result in a decrease in the number of receptors.

Receptor sensitization or upregulation
The process by which the concentration and affinity of receptors are increased. Decreased exposure to an agonist can result in an increase in the number of receptors.

Retrograde transmission
Movement of the endocannabinoid (AEA or 2-AG) from the postsynaptic neuron to the presynaptic neuron.

Retrospective study
A study in which individuals are sampled and information is collected about their past.

Synergistic
The interaction of two or more compounds, when their combined effect is greater than the sum of the effects seen when each compound is given alone.

Synapse
The site of transmission of electric nerve impulses between two nerve cells (neurons) or between a neuron and a gland or muscle cell (effector).

Titration
Adjusting the dose for maximum effect.

Transdermal
An application to the skin.

Transient receptor potential V (TRPV1)
Also known as the capsaicin receptor and the vanilloid receptor 1, it is a receptor found within the cell that acts as an ion channel upon which anandamide acts. It is important in the detection and regulation of body temperature and pain.

Vagus nerve
The tenth cranial nerve that carries information between the autonomic nervous system and the brainstem.

Vaporization
The process of heating cannabis until violative active cannabinoids are vaporized.

Chapter 1

1. Mechoulam et al., "Early Phytocannabinoid Chemistry."
2. ElSohly et al., "Changes in Cannabis Potency."
3. Pisanti and Bifulco, "Medical Cannabis."
4. Pisanti and Bifulco.
5. Mechoulam et al., "Early Phytocannabinoid Chemistry."
6. O'Shaughnessy, "On the Preparations of Indian Hemp."
7. Abel, "Marihuana."
8. Abel.
9. Mechoulam and Carlini, "Toward Drugs Derived from Cannabis"; Cunha et al., "Chronic Administration of Cannabidiol."
10. Mead, "International Control of Cannabis."
11. Haney, "Perspectives on Cannabis Research."
12. Haney, "Perspectives on Cannabis Research."
13. Haney, "Perspectives on Cannabis Research."

Chapter 2

1. Barlow et al., "XL.—Charas."
2. Gaoni and Mechoulam, "Isolation, Structure, and Partial Synthesis."
3. Devane et al., "Determination and Characterization."
4. Munro, and Abu-Shaar, "Molecular Characterization."
5. Herkenham et al., "Cannabinoid Receptor Localization."
6. Hughes et al., "Identification of Two Related Pentapeptides."
7. Devane et al., "Isolation and Structure."
8. Mechoulam et al., "Identification of an Endogenous 2-Monoglyceride."
9. Sugiura et al., "2-Arachidonoylglycerol."
10. Pacher and Mechoulam, "Is Lipid Signaling?"
11. Pacher and Mechoulam, "Is Lipid Signaling?"
12. Laprairie et al., "Cannabidiol Is a Negative Allosteric Modulator." nm

Chapter 3

1. For an excellent review, see Morales et al., "Molecular Targets of the Phytocannabinoids."
2. Russo, "Taming THC."

3. Gaoni and Mechoulam, "Isolation, Structure, and Partial Synthesis."
4. Martin et al., "Behavioral, Biochemical, and Molecular modeling."
5. Abrahamov et al. "Efficient New Cannabinoid Antiemetic."
6. Wood et al., "XL.—Charas."; Cahn, "174. Cannabis indica resin."
7. Adams et al., "Structure of Cannabidiol."
8. Mechoulam and Shvo, "Hashish."
9. Parker et al., *CBD*.
10. Laprairie et al., "Cannabidiol Is a Negative Allosteric Modulator."
11. ElSohly et al., "Comprehensive Review of Cannabis Potency."
12. Bonn-Miller et al., "Labeling Accuracy of Cannabidiol Extracts."
13. Cuttler et al., "Acute Effects of Cannabigerol."
14. Russo, "Taming THC."
15. Hazekamp et al., "Medicinal Use of Cannabis."
16. Huestis et al., "Estimating the Time."
17. Huestis and Smith, "Cannabinoid Pharmacokinetics."
18. Huestis and Smith.
19. Huestis and Smith.
20. Birnbaum et al., "Food Effect on Pharmacokinetics."
21. Patrician et al., "Examination of a New Delivery Approach"; Atsmon et al., "Single-Dose Pharmacokinetics."
22. Jones and Pertwee, "Metabolic Interaction in Vivo."
23. Geffrey et al., "Drug–Drug Interaction."
24. Gaston et al., "Drug–Drug Interactions."
25. Balachandran et al., "Cannabidiol Interactions with Medications."
26. Balachandran et al. "Cannabidiol Interactions with Medications."
27. Chesney et al., "Adverse Effects of Cannabidiol."
28. Bonn-Miller et al., "Labeling Accuracy of Cannabidiol Extracts."
29. Englund et al., "Cannabidiol Inhibits THC-Elicited Paranoid Symptoms"; Morgan et al., "Individual and Combined Effects."
30. Laprairie et al., "Cannabidiol Is a Negative Allosteric Modulator."
31. Chandra et al., "New Trends in Cannabis Potency."
32. Solowij et al., "Therapeutic Effects of Prolonged Cannabidiol Treatment."
33. King et al., "Single and Combined Effects"; Rock and Parker, "Synergy Between Cannabidiol, Cannabidiolic Acid"; Rock et al., "Effect of Combined Doses."
34. Jones and Pertwee, "Metabolic Interaction in Vivo."
35. King et al., "Single and Combined Effects."
36. Russo and Guy, "Tale of Two Cannabinoids."

37. Schoedel et al., "Randomized, Double-Blind, Placebo-Controlled, Cross-over Study."

38. Winhusen et al., "Regular Cannabis Use."

39. Gieringer et al., "Cannabis Vaporizer Combines Efficient Delivery."

40. Mittleman et al., "Triggering Myocardial Infarction by Marijuana."

41. Hartman and Huestis, "Cannabis Effects on Driving Skills."

Chapter 4

1. For a more detailed account of this research see Parker, *Cannabinoids and the Brain* (MIT Press, 2017).

2. Connor et al., "Cannabis Use and Cannabis Use Disorder."

3. Piomelli et al. "Legal or Illegal."

4. Haney et al., "Nabilone Decreases Marijuana Withdrawal."

5. Freels et al., "Vaporized Cannabis Extracts."

6. Parsons and Hurd, "Endocannabinoid Signalling in Reward and Addiction."

7. Pertwee, "Inverse Agonism and Neutral Antagonism."

8. Soler-Cedeño et al., "AM6527, a Neutral CB1 Receptor Antagonist."

9. Ren et al., "Cannabidiol, a Nonpsychotropic Component."

10. Renard et al., "Neuronal and Molecular Effects."

11. Scicluna et al., "Cannabidiol Reduced the Severity of Gastrointestinal Symptoms."

12. Ren et al., "Cannabidiol, a Nonpsychotropic Component."

13. Hurd et al., "Cannabidiol for the Reduction of Cue-Induced Craving."

14. Manini et al., "Safety and Ppharmacokinetics of Oral Cannabidiol."

15. Hurd et al., "Early Phase in the Development of Cannabidiol."

16. Hurd et al., "Cannabidiol for the Reduction of Cue-Induced Craving."

Chapter 5

1. Hill et al., "Functional Interactions."

2. For an excellent review, see Patel et al.,"Effects of Phytocannabinoids."

3. Hindocha et al., "Acute Effects of Delta-9-Tetrahydrocannabinol"; Berga-maschi et al., "Cannabidiol Reduces the Anxiety."

4. Fraser, "Use of a Synthetic Cannabinoid."

5. Jetly et al., "Efficacy of Nabilone."

6. Fraser, "Use of a Synthetic Cannabinoid."

7. Das et al., "Cannabidiol Enhances Consolidation."

8. Patel et al, "Effects of Phytocannabinoids."

9. Zuardi et al., "Action of Cannabidiol."

10. Cuttler et al., "Acute Effects of Cannabigerol."

11. Fuss et al., "Runner's High Depends on Cannabinoid Receptors."

12. Patel et al., "Effects of Phytocannabinoids."

13. Bluett et al., "Central Anandamide Deficiency."

14. Marsicano et al., "Endogenous Cannabinoid System."

15. Hill and Tasker, "Endocannabinoid Signaling."

16. Morena et al., "Neurobiological Interactions."

17. Gates et al., "Effects of Cannabinoid Administration on Sleep."

18. Gates et al., "Effects of Cannabinoid Administration on Sleep."

19. Fraser, "Use of a Synthetic Cannabinoid."

20. AminiLari et al., "Medical Ccannabis and Cannabinoids."

21. Miranda et al., "Role of Cannabis."

22. Ranum et al., "Use of Cannabidiol."

23. J. J. Moreau, *Hashish and Mental Illness*, translated from French.by Gordon J. Barnett (New York: Raven, 1973 [1845]).

24. Colizzi et al., "Unraveling the Intoxicating and Therapeutic Effects."

25. Hindley et al., "Psychiatric Symptoms Caused by Cannabis Constituents."

26. Arseneault et al., "Causal Association Between Cannabis and Psychosis."

27. Moore et al., "Cannabis Use and Risk."; Large et al., "Cannabis Use and Earlier Onset."

28. Gage et al., "Cannabis and Psychosis."

29. Wilkinson et al., "Impact of Cannabis Use."

30. Henquet et al., "Gene-Environment Interplay."

31. Caspi et al., "Moderation of the Effect."; Costas et al., "Interaction Between COMT Haplotypes and Cannabis"; Zammit et al., "Cannabis, COMT and Psychotic Experiences."

32. Costas et al., "Interaction Between COMT Haplotypes and Cannabis."

33. Caspi et al., "Moderation of the Effect."

34. Power et al., "Genetic Predisposition to Schizophrenia."

35. Cheng et al., "Relationship Between Cannabis Use, schizophrenia."; Pasman et al., "GWAS of Lifetime Cannabis Use"; Gage et al., "Assessing Causality in Associations."

36. D'Souza et al., "Consensus Paper of the WFSBP Task Force."

37. Leweke et al., "Cannabidiol Enhances Anandamide Signaling."

38. McGuire et al., "Cannabidiol (CBD) as an Adjunctive Therapy."

39. Bhattacharyya et al., "Effect of Cannabidiol on Medial Temporal."

40. Boggs et al., " Effects of cannabidiol (CBD) on cognition."

41. Zuardi et al., "Cannabidiol for the Treatment of Psychosis."

Chapter 6

1. For a review, see Marsicano and Lafenêtre, "Roles of the Endocannabinoid Ssystem."
2. Heifets and Castillo, "Endocannabinoid Signaling."
3. Campolongo et al., "Endocannabinoids in the Rat Basolateral Amygdala."
4. Curran et al., "Cognitive and Subjective Dose-Response Effects."
5. Morgan et al., "Individual and Combined Effects."
6. Broyd et al., "Acute and Chronic Effects of Cannabinoids."
7. Meier et al., "Persistent Cannabis Users Show Neuropsychological Decline."
8. Mokrysz et al., "Are IQ and Educational Outcomes in Teenagers?"; Jackson et al., "Impact of Adolescent Marijuana Use."
9. Meier et al., "Persistent Cannabis Users Show Neuropsychological Decline"; Curran et al., "Keep Off the Grass?"
10. Curran et al., "Keep Off the Grass?"
11. Schreiner and Dunn, "Residual Effects of Cannabis Use."
12. Curran et al., "Keep Off the Grass?"
13. Hirvonen et al., "Reversible and Regionally Selective Downregulation."
14. D'Souza et al., "Rapid Changes in CB1 Receptor Availability."
15. Sapolsky, *Behave*.
16. B. Gunasekera, K. Diederen, and S. Bhattacharyya, "Cannabinoids, Reward Processing, and Psychosis," *Psychopharmacology (Berlin)* 239, no. 5 (May 2022): 1157–1177. https://doi.org/10.1007/s00213-021-05801-2.
17. Weiland et al., "Daily Marijuana Use."
18. Weiland et al., "Daily Marijuana Use."
19. Gunasekera et al., "Cannabinoids, Reward Processing, and Psychosis."
20. Harper et al., "Effects of Alcohol and Cannabis Use."

Chapter 7

1. Hollister, "Actions of Various Marihuana Derivatives."
2. Monteleone et al., "Blood Levels of the Endocannabinoid Anandamide."
3. Després et al., "Effects of Rimonabant."; Pi-Sunyer et al., "Effect of Rimonabant"; Scheen et al., "Efficacy and Tolerability of Rimonabant"; Van Gaal et al., "Effects of the Cannabinoid-1 Receptor Blocker."
4. Tam et al., "Peripheral Cannabinoid-1 Receptor Inverse Agonism."
5. Bisogno et al., "Novel Fluorophosphonate Inhibitor."
6. Jung et al., "2-Arachidonoylglycerol Signaling in Forebrain."
7. Horswill et al., "PSNCBAM-1, a Novel Allosteric Antagonist."
8. Deveaux et al., "Cannabinoid CB2 Receptor."
9. Thomas et al., "Evidence That the Plant Cannabinoid."

Chapter 8

1. Rock and Parker, "Cannabinoids as Potential Treatment."
2. Meiri et al., "Efficacy of Dronabinol."
3. Meiri et al., "Efficacy of Dronabinol alone."
4. Rock et al., "Cannabidiol, a Non-Psychotropic Component."
5. For a review, see Rock et al., "Therapeutic Potential of Cannabidiol."
6. Dall'Stella et al., "Case Report."
7. Rock and Parker, "Cannabinoids as Potential Treatment."
8. Sticht et al., "Endocannabinoid Regulation of Nausea is Mediate by 2-Arachidonoylglycerol (2-AG) in the Rat Visceral Insular Cortex."
9. Napadow et al., "Brain Circuitry Underlying the Temporal Evolution."
10. Rock et al., "Cannabidiol, a Non-Psychotropic Component."
11. Limebeer et al., "Nausea-Induced 5-HT Release."; Sticht et al., "Endocannabinoid Regulation of Nausea."
12. DeVuono et al., "Endocannabinoid Signaling in Stress."
13. DeVuono et al., "Conditioned Gaping."; DeVuono and Parker, "Cannabinoid Hyperemesis Syndrome."
14. DeVuono et al., "Cannabidiol Interferes with Establishment."

Chapter 9

1. Cuttler et al., "Large-Scale Naturalistic Examination."
2. Vigil et al., "Associations Between Medical Cannabis and Prescription Opioid Use."
3. Costa and Comelli, "Pain."
4. Costa and Comelli.
5. Ward et al., "Cannabidiol Inhibits Paclitaxel-Induced Neuropathic Pain."
6. Varvel et al., "Interactions Between THC and Cannabidiol."
7. Portenoy et al., "Nabiximols for Opioid-Treated Cancer Patients."
8. Ignatowska-Jankowska et al., "Cannabinoid CB1 Receptor-Positive Allosteric Modulator."
9. Fowler, "The Potential of Inhibitors of Endocannabinoid Metabolism."
10. J. Guindon and A. G. Hohmann, "Cannabinoid 2 Receptors: A Therapeutic Target for the Treatment of Inflammatory and Neuropathic Pain," *British Journal of Pharmacology* 153, no. 2 (2008): 319–334.
11. Wilsey et al., "Low-Dose Vaporized Cannabis."
12. Fisher et al., "Cannabinoids, Cannabis, and Cannabis-Based Medicine."
13. Fisher et al., "Cannabinoids, Cannabis, and Cannabis-Based Medicine."
14. Whiting et al., "Cannabinoids for Medical Use"; Stockings et al., "Cannabis and Cannabinoids"; Fisher et al., "Cannabinoids, Cannabis, and Cannabis-Based Medicine"; Bilbao and Spanagel, "Medical Cannabinoids."

15. Lötsch et al., "Current Evidence of Cannabinoid-Based Analgesia."
16. Häuser et al., "Cannabis-Based Medicines."
17. Bhaskar et al., "Consensus Recommendations on Dosing."

Chapter 10
1. Cristino et al., "Cannabinoids and the Expanded Endocannabinoid System."
2. Williams et al., "Cannabis and Epilepsy."
3. Soltesz et al., "Weeding Out Bad Waves."
4. Russo, "Cannabis and Epilepsy."
5. Shaughnessy, "On the Preparations of Indian Hemp."
6. Devinsky et al., "Cannabidiol."
7. Mechoulam and Carlini, "Toward Drugs Derived from Cannabis."
8. Cunha et al., "Chronic Administration of Cannabidiol."
9. Mechoulam et al., "Early Phytocannabinoid Chemistry."
10. Whiting et al., "Cannabinoids for Medical Use."
11. Devinsky et al., "Trial of Cannabidiol for Drug-Resistant Seizures."
12. Thiele et al., "Cannabidiol in Patients with Seizures."
13. Arzimanoglou et al., "Epilepsy and Cannabidiol."; Geffrey et al., "Drug–Drug Interaction."
14. Gaston et al., "Drug–Drug Interactions."
15. Balachandran et al. "Cannabidiol Interactions with Medications."
16. Ruwini et al., "Current Aspects of the Endocannabinoid System."
17. Fernández-Ruiz et al., "Neurodegenerative Disorders Other Than Multiple Sclerosis."
18. Cristino et al., "Cannabinoids and the Expanded Endocannabinoid System."
19. Paes-Colli et al., "Phytocannabinoids and Cannabis-Based Products."
20. For a review, see Pacher and Mechoulam, "Is lipid signaling?"
21. Fernández-Ruiz et al., "Neurodegenerative Disorders Other Than Multiple Sclerosis."
22. Fernández-Ruiz et al., "Neurodegenerative Disorders Other Than Multiple Sclerosis."
23. Kuharic et al., "Cannabinoids for the Treatment of Dementia."
24. Curtis et al., "Pilot Study Using Nabilone."
25. Fernández-Ruiz et al., "Neurodegenerative Disorders Other Than Multiple Sclerosis."
26. Urbi et al., "Effects of Cannabis in Parkinson's Disease."
27. De Faria et al., "Effects of Acute Cannabidiol Administration."
28. Pryce and Baker, "Cannabis and Multiple Sclerosis."

29. Pryce and Baker.
30. Whiting et al., "Cannabinoids for Medical Use."
31. Pryce and Baker, "Cannabis and Multiple Sclerosis."
32. Zajicek et al., "Effect of Dronabinol on Progression."
33. Pryce and Baker, "Cannabis and Multiple Sclerosis."

Chapter 11

1. Maccarrone et al., "Goods and Bads of the Endocannabinoid System."
2. Maccarrone et al., "Goods and Bads of the Endocannabinoid System."
3. Haney et al., "Nabilone Decreases Marijuana Withdrawal."
4. Paul et al., "Associations Between Prenatal Cannabis Exposure."
5. Bilbao and Spanagel, "Medical cannabinoids."
6. Bilbao and Spanagel, "Medical cannabinoids"; Grimison et al., "Oral THC"; Duran et al., "Preliminary Efficacy and Safety."
7. Spanagel and Bilbao, "Approved Cannabinoids for Medical Purposes."
8. Whiting et al., "Cannabinoids for Medical Us."
9. Amato et al., "Patent Update on Cannabinoid Receptor 1 Antagonists."
10. Ignatowska-Jankowska et al., "Cannabinoid CB1 Receptor-Positive Allosteric Modulator."
11. Cinar et al., "Therapeutic Potential."; Sink et al., "Novel Cannabinoid CB1 Receptor Neutral Antagonist."; Cluny et al., "Novel Peripherally Restricted Cannabinoid Receptor Antagonist."
12. Pacher and Mechoulam, "Is Lipid Signaling?"
13. Guindon and Hohmann, "Cannabinoid CB2 Receptors."
14. Maccarrone et al., "Goods and Bads of the Endocannabinoid System."
15. Brennecke et al., "Cannabinoid Receptor Type 2 Ligands"; Maccarrone et al., "Goods and Bads of the Endocannabinoid System."
16. Justinova et al., "Fatty Acid Amide Hydrolase Inhibition"; Maccarrone et al., "Goods and Bads of the Endocannabinoid System."
17. Schmidt et al., "Effects of Inhibition."
18. D'Souza et al., "Efficacy and Safety."

BIBLIOGRAPHY

Abel, Ernest L. "Marihuana." In *Psychoactive Drugs and Sex*, 119–168. Springer, 1980.

Abrahamov, Aya, Avraham Abrahamov, and R. Mechoulam. "An Efficient New Cannabinoid Antiemetic in Pediatric Oncology." *Life Sciences* 56, no. 23 (May 5, 1995): 2097–2102. https://doi.org/10.1016/0024-3205(95)00194-B. www.sciencedirect.com/science/article/pii/002432059500194B.

Adams, Roger, Madison Hunt, and J. H. Clark. "Structure of Cannabidiol, a Product Isolated from the Marihuana Extract of Minnesota Wild Hemp. I." *Journal of the American Chemical Society* 62, no. 1 (January 1, 1940): 196–200. https://doi.org/10.1021/ja01858a058. https://doi.org/10.1021/ja01858a058.

Amato, G., N. S. Khan, and R. Maitra. "A Patent Update on Cannabinoid Receptor 1 Antagonists (2015–2018)." *Expert Opinion on Therapeutic Patents* 29, no. 4 (April 2019): 261–269. https://doi.org/10.1080/13543776.2019.1597851.

AminiLari, M., L. Wang, S. Neumark, T. Adli, R. J. Couban, A. Giangregorio, C. E. Carney, and J. W. Busse. "Medical Cannabis and Cannabinoids for Impaired Sleep: A Systematic Review and Meta-Analysis of Randomized Clinical Trials." *Sleep* 45, no. 2 (February 14, 2022). https://doi.org/10.1093/sleep/zsab234.

Arseneault, L., M. Cannon, J. Witton, and R. M. Murray. "Causal Association Between Cannabis and Psychosis: Examination of the Evidence." *British Journal of Psychiatry* 184 (February 2004): 110–117. https://doi.org/10.1192/bjp.184.2.110.

Arzimanoglou, A., U. Brandl, J. H. Cross, A. Gil-Nagel, L. Lagae, C. J. Landmark, N. Specchio, et al. "Epilepsy and Cannabidiol: A Guide to Treatment." *Epileptic Disorders* 22, no. 1 (February 1, 2020): 1–14. https://doi.org/10.1684/epd.2020.1141.

Atsmon, Jacob, Daphna Heffetz, Lisa Deutsch, Frederic Deutsch, and Hagit Sacks. "Single-Dose Pharmacokinetics of Oral Cannabidiol Following Administration of Ptl101: A New Formulation Based on Gelatin Matrix Pellets Technology." *Clinical Pharmacology in Drug Development* 7, no. 7 (2018): 751–758.

Balachandran, P., M. Elsohly, and K. P. Hill. "Cannabidiol Interactions with Medications, Illicit Substances, and Alcohol: A Comprehensive Review." *Journal of General Internal Medicine* (January 29, 2021). https://doi.org/10.1007/s11606-020-06504-8.

Bergamaschi, M. M., R. H. Queiroz, M. H. Chagas, D. C. de Oliveira, B. S. De Martinis, F. Kapczinski, J. Quevedo, et al. "Cannabidiol Reduces the Anxiety Induced by Simulated Public Speaking in Treatment-Naive Social Phobia Patients." *Neuropsychopharmacology* 36, no. 6 (May 2011): 1219–1226. https://doi.org/10.1038/npp.2011.6.

Bhaskar, A., A. Bell, M. Boivin, W. Briques, M. Brown, H. Clarke, C. Cyr, et al. "Consensus Recommendations on Dosing and Administration of Medical Cannabis to Treat Chronic Pain: Results of a Modified Delphi Process." *Journal of Cannabis Research* 3, no. 1 (July 2, 2021): 22. https://doi.org/10.1186/s42238-021-00073-1.

Bhattacharyya, S., R. Wilson, E. Appiah-Kusi, A. O'Neill, M. Brammer, J. Perez, R. Murray, et al. "Effect of Cannabidiol on Medial Temporal, Midbrain, and Striatal Dysfunction in People at Clinical High Risk of Psychosis: A Randomized Clinical Trial." *JAMA Psychiatry* 75, no. 11 (November 1, 2018): 1107–1117. https://doi.org/10.1001/jamapsychiatry.2018.2309.

Bilbao, A., and R. Spanagel. "Medical Cannabinoids: A Pharmacology-Based Systematic Review and Meta-Analysis for All Relevant Medical Indications." *BMC Medicine* 20, no. 1 (August 19, 2022): 259. https://doi.org/10.1186/s12916-022-02459-1.

Birnbaum, A. K., A. Karanam, S. E. Marino, C. M. Barkley, R. P. Remmel, M. Roslawski, M. Gramling-Aden, and I. E. Leppik. "Food Effect on Pharmacokinetics of Cannabidiol Oral Capsules in Adult Patients with Refractory Epilepsy." *Epilepsia* 60, no. 8 (August 2019): 1586–1592. https://doi.org/10.1111/epi.16093.

Bisogno, T., A. Mahadevan, R. Coccurello, J. W. Chang, M. Allarà, Y. Chen, G. Giacovazzo, et al. "A Novel Fluorophosphonate Inhibitor of the Biosynthesis of the Endocannabinoid 2-Arachidonoylglycerol with Potential Anti-Obesity Effects." *British Journal of Pharmacology* 169, no. 4 (June 2013): 784–793. https://doi.org/10.1111/bph.12013.

Bluett, R. J., J. C. Gamble-George, D. J. Hermanson, N. D. Hartley, L. J. Marnett, and S. Patel. "Central Anandamide Deficiency Predicts Stress-Induced Anxiety: Behavioral Reversal Through Endocannabinoid Augmentation." *Translational Psychiatry* 4 (July 8, 2014): e408. https://doi.org/10.1038/tp.2014.53.

Boggs, D. L., T. Surti, A. Gupta, S. Gupta, M. Niciu, B. Pittman, A. M. Schnakenberg Martin, et al. "The Effects of Cannabidiol (CBD) on Cognition and Symptoms in Outpatients with Chronic Schizophrenia a Randomized Placebo Controlled Trial." *Psychopharmacology (Berlin)* 235, no. 7 (July 2018): 1923–1932. https://doi.org/10.1007/s00213-018-4885-9.

Bonn-Miller, M. O., M. J. E. Loflin, B. F. Thomas, J. P. Marcu, T. Hyke, and R. Vandrey. "Labeling Accuracy of Cannabidiol Extracts Sold Online." *JAMA* 318, no. 17 (November 7, 2017): 1708–1709. https://doi.org/10.1001/jama.2017.11909.

Bosnjak Kuharic, D., D. Markovic, T. Brkovic, M. Jeric Kegalj, Z. Rubic, A. Vuica Vukasovic, A. Jeroncic, and L. Puljak. "Cannabinoids for the Treatment of Dementia." *Cochrane Database System Review* 9, no. 9 (September 17, 2021): Cd012820. https://doi.org/10.1002/14651858.CD012820.pub2.

Brennecke, Benjamin, Thais Gazzi, Kenneth Atz, Jürgen Fingerle, Pascal Kuner, Torsten Schindler, Guy de Weck, Marc Nazaré, and Uwe Grether. "Cannabinoid Receptor Type 2 Ligands: An Analysis of Granted Patents Since 2010." *Pharmaceutical Patent Analyst* 10, no. 3 (2021): 111–163.

Broyd, S. J., H. H. van Hell, C. Beale, M. Yücel, and N. Solowij. "Acute and Chronic Effects of Cannabinoids on Human Cognition—A Systematic Review." *Biological Psychiatry* 79, no. 7 (April 1, 2016): 557–567. https://doi.org/10.1016/j.biopsych.2015.12.002.

Cahn, Robert Sidney. "174. Cannabis Indica Resin. Part III. The Constitution of Cannabinol." *Journal of the Chemical Society (Resumed)* (1932): 1342–1353.

Campolongo, P., B. Roozendaal, V. Trezza, D. Hauer, G. Schelling, J. L. McGaugh, and V. Cuomo. "Endocannabinoids in the Rat Basolateral Amygdala Enhance Memory Consolidation and Enable Glucocorticoid Modulation of Memory." *Proceedings of the National Academy of Sciences USA* 106, no. 12 (March 24, 2009): 4888–4893. https://doi.org/10.1073/pnas.0900835106.

Caspi, A., T. E. Moffitt, M. Cannon, J. McClay, R. Murray, H. Harrington, A. Taylor, et al. "Moderation of the Effect of Adolescent-Onset Cannabis Use on Adult Psychosis by a Functional Polymorphism in the Catechol-O-Methyltransferase Gene: Longitudinal Evidence of a Gene X Environment Interaction." *Biological Psychiatry* 57, no. 10 (May 15, 2005): 1117–1127. https://doi.org/10.1016/j.biopsych.2005.01.026.

Chandra, Suman, Mohamed M. Radwan, Chandrani G. Majumdar, James C. Church, Tom P. Freeman, and Mahmoud A. ElSohly. "New Trends in Cannabis

Potency in USA and Europe During the Last Decade (2008–2017)." *European Archives of Psychiatry and Clinical Neuroscience* 269 (2019): 5–15.

Cheng, W., N. Parker, N. Karadag, E. Koch, G. Hindley, R. Icick, A. Shadrin, et al. "The Relationship Between Cannabis Use, Schizophrenia, and Bipolar Disorder: A Genetically Informed Study." *Lancet Psychiatry* 10, no. 6 (June 2023): 441–451. https://doi.org/10.1016/s2215-0366(23)00143-8.

Chesney, E., D. Oliver, A. Green, S. Sovi, J. Wilson, A. Englund, T. P. Freeman, and P. McGuire. "Adverse Effects of Cannabidiol: A Systematic Review and Meta-Analysis of Randomized Clinical Trials." *Neuropsychopharmacology* (April 8, 2020). https://doi.org/10.1038/s41386-020-0667-2.

Cinar, R., M. R. Iyer, and G. Kunos. "The Therapeutic Potential of Second and Third Generation CB_1R Antagonists." *Pharmacology and Theraputics* 208 (April 2020): 107477. https://doi.org/10.1016/j.pharmthera.2020.107477.

Cluny, N. L., V. K. Vemuri, A. P. Chambers, C. L. Limebeer, H. Bedard, J. T. Wood, B. Lutz, et al. "A Novel Peripherally Restricted Cannabinoid Receptor Antagonist, Am6545, Reduces Food Intake and Body Weight, but Does Not Cause Malaise, in Rodents." *British Journal of Pharmacology* 161, no. 3 (2010): 629–642. https://doi.org/10.1111/j.1476-5381.2010.00908.x.

Colizzi, Marco, Mirella Ruggeri, and Sagnik Bhattacharyya. "Unraveling the Intoxicating and Therapeutic Effects of Cannabis Ingredients on Psychosis and Cognition." *Frontiers in Psychology* 11 (2020): 833.

Connor, Jason P., Daniel Stjepanović, Bernard Le Foll, Eva Hoch, Alan J. Budney, and Wayne D. Hall. "Cannabis Use and Cannabis Use Disorder." *Nature Reviews Disease Primers* 7, no. 1 (2021): 16.

Cooray, Ruwini, Veer Gupta, and Cenk Suphioglu. "Current Aspects of the Endocannabinoid System and Targeted THC and CBD Phytocannabinoids as Potential Therapeutics for Parkinson's and Alzheimer's Diseases: A Review." *Molecular Neurobiology* 57, no. 11 (2020): 4878–4890.

Costa, Barbara, and Francesca Comelli. "Pain." In *Handbook of Cannabis*, edited by Roger Pertwee, 473–486. Oxford University Press, 2014.

Costas, J., J. Sanjuán, R. Ramos-Ríos, E. Paz, S. Agra, A. Tolosa, M. Páramo, J. Brenlla, and M. Arrojo. "Interaction Between COMT Haplotypes and Cannabis in Schizophrenia: A Case-Only Study in Two Samples from Spain." *Schizophrenia Research* 127, nos. 1–3 (April 2011): 22–27. https://doi.org/10.1016/j.schres.2011.01.014.

Cristino, L., T. Bisogno, and V. Di Marzo. "Cannabinoids and the Expanded Endocannabinoid System in Neurological Disorders." *Nature Reviews Neurology* 16, no. 1 (January 2020): 9–29. https://doi.org/10.1038/s41582-019-0284-z.

Cunha, J. M., E. A. Carlini, A. E. Pereira, O. L. Ramos, C. Pimentel, R. Gagliardi, W. L. Sanvito, N. Lander, and R. Mechoulam. "Chronic Administration of Cannabidiol to Healthy Volunteers and Epileptic Patients." *Pharmacology* 21, no. 3 (1980): 175–185.

Curran, H. V., C. Brignell, S. Fletcher, P. Middleton, and J. Henry. "Cognitive and Subjective Dose-Response Effects of Acute Oral Delta 9-Tetrahydrocannabinol (THC) in Infrequent Cannabis Users." *Psychopharmacology (Berlin)* 164, no. 1 (October 2002): 61–70. https://doi.org/10.1007/s00213-002-1169-0.

Curran, H. V., T. P. Freeman, C. Mokrysz, D. A. Lewis, C. J. Morgan, and L. H. Parsons. "Keep Off the Grass? Cannabis, Cognition and Addiction." *Nature Reviews Neuroscience* 17, no. 5 (May 2016): 293–306. https://doi.org/10.1038/nrn.2016.28.

Curtis, Adrienne, Ian Mitchell, Smitaa Patel, Natalie Ives, and Hugh Rickards. "A Pilot Study Using Nabilone for Symptomatic Treatment in Huntington's Disease." *Movement Disorders: Official Journal of the Movement Disorder Society* 24, no. 15 (2009): 2254–2259.

Cuttler, C., E. M. LaFrance, and R. M. Craft. "A Large-Scale Naturalistic Examination of the Acute Effects of Cannabis on Pain." *Cannabis and Cannabinoid Research* (October 23, 2020). https://doi.org/10.1089/can.2020.0068.

Cuttler, Carrie, Amanda Stueber, Ziva D. Cooper, and Ethan Russo. "Acute Effects of Cannabigerol on Anxiety, Stress, and Mood: A Double-Blind, Placebo-Controlled, Crossover, Field Trial." *Scientific Reports* 14, no. 1 (2024): 16163.

D'Souza, D. C., J. A. Cortes-Briones, M. Ranganathan, H. Thurnauer, G. Creatura, T. Surti, B. Planeta, et al. "Rapid Changes in CB1 Receptor Availability in Cannabis Dependent Males After Abstinence from Cannabis." *Biological Psychiatry: Cognitive Neuroscience and Neuroimaging* 1, no. 1 (January 1, 2016): 60–67. https://doi.org/10.1016/j.bpsc.2015.09.008.

D'Souza, D. C., J. Cortes-Briones, G. Creatura, G. Bluez, H. Thurnauer, E. Deaso, K. Bielen, et al. "Efficacy and Safety of a Fatty Acid Amide Hydrolase Inhibitor (Pf-04457845) in the Treatment of Cannabis Withdrawal and Dependence in Men: A Double-Blind, Placebo-Controlled, Parallel Group, Phase 2a Single-Site

Randomised Controlled Trial." *Lancet Psychiatry* 6, no. 1 (January 2019): 35–45. https://doi.org/10.1016/s2215-0366(18)30427-9.

D'Souza, Deepak Cyril, Marta DiForti, Suhas Ganesh, Tony P. George, Wayne Hall, Carsten Hjorthøj, Oliver Howes, et al. "Consensus Paper of the WFSBP Task Force on Cannabis, Cannabinoids and Psychosis." *World Journal of Biological Psychiatry* 23, no. 10 (2022): 719–742.

Dall'Stella, Paula B., Marcos F. L. Docema, Marcos V. C. Maldaun, Olavo Feher, and Carmen L. P. Lancellotti. "Case Report: Clinical Outcome and Image Response of Two Patients with Secondary High-Grade Glioma Treated with Chemoradiation, PCV, and Cannabidiol." Case Report. *Frontiers in Oncology* 8, no. 643 (January 18, 2019). https://doi.org/10.3389/fonc.2018.00643. www.frontiersin.org/article/10.3389/fonc.2018.00643.

Das, R. K., S. K. Kamboj, M. Ramadas, K. Yogan, V. Gupta, E. Redman, H. V. Curran, and C. J. Morgan. "Cannabidiol Enhances Consolidation of Explicit Fear Extinction in Humans." *Psychopharmacology (Berlin)* 226, no. 4 (April 2013): 781–792. https://doi.org/10.1007/s00213-012-2955-y.

De Faria, Stephanie Martins, Daiene de Morais Fabrício, Vitor Tumas, Paula Costa Castro, Moacir Antonelli Ponti, Jaime E. C. Hallak, Antonio W. Zuardi, José Alexandre S. Crippa, and Marcos Hortes Nisihara Chagas. "Effects of Acute Cannabidiol Administration on Anxiety and Tremors Induced by a Simulated Public Speaking Test in Patients with Parkinson's Disease." *Journal of Psychopharmacology* 34, no. 2 (2020): 189–196.

Després, J. P., A. Golay, and L. Sjöström. "Effects of Rimonabant on Metabolic Risk Factors in Overweight Patients with Dyslipidemia." *New England Journal of Medicine* 353, no. 20 (November 17, 2005): 2121–2134. https://doi.org/10.1056/NEJMoa044537.

Devane, W. A., F. A. Dysarz, III, M. R. Johnson, L. S. Melvin, and A. C. Howlett. "Determination and Characterization of a Cannabinoid Receptor in Rat Brain." *Molecular Pharmacology* 34, no. 5 (November 1988): 605–613.

Devane, W. A., L. Hanus, A. Breuer, R. G. Pertwee, L. A. Stevenson, G. Griffin, D. Gibson, et al. "Isolation and Structure of a Brain Constituent That Binds to the Cannabinoid Receptor." *Science (New York, N.Y.)* 258, no. 5090 (1992): 1946–1949. https://doi.org/10.1126/science.1470919.

Deveaux, V., T. Cadoudal, Y. Ichigotani, F. Teixeira-Clerc, A. Louvet, S. Manin, J. T. Nhieu, et al. "Cannabinoid CB2 Receptor Potentiates Obesity-Associated

Inflammation, Insulin Resistance and Hepatic Steatosis." *PloS One* 4, no. 6 (June 9 2009): e5844. https://doi.org/10.1371/journal.pone.0005844.

Devinsky, O., J. H. Cross, L. Laux, E. Marsh, I. Miller, R. Nabbout, I. E. Scheffer, E. A. Thiele, and S. Wright. "Trial of Cannabidiol for Drug-Resistant Seizures in the Dravet Syndrome." *New England Journal of Medicine* 376, no. 21 (May 25, 2017): 2011–2020. https://doi.org/10.1056/NEJMoa1611618.

Devinsky, O., M. R. Cilio, H. Cross, J. Fernández-Ruiz, J. French, C. Hill, R. Katz, et al. "Cannabidiol: Pharmacology and Potential Therapeutic Role in Epilepsy and Other Neuropsychiatric Disorders." *Epilepsia* 55, no. 6 (June 2014): 791–802. https://doi.org/10.1111/epi.12631.

DeVuono, M. V., K. M. Hrelja, L. Sabaziotis, A. Rajna, E. M. Rock, C. L. Limebeer, D. M. Mutch, and L. A. Parker. "Conditioned Gaping Produced by High Dose $\Delta(9)$-Tetrahydracannabinol: Dysregulation of the Hypothalamic Endocannabinoid System." *Neuropharmacology* 141 (October 2018): 272–282. https://doi.org/10.1016/j.neuropharm.2018.08.039.

DeVuono, M. V., T. Venkatesan, and C. J. Hillard. "Endocannabinoid Signaling in Stress, Nausea, and Vomiting." *Neurogastroenterology and Motility* (September 2, 2024): e14911. https://doi.org/10.1111/nmo.14911.

DeVuono, Marieka V., and Linda A. Parker. "Cannabinoid Hyperemesis Syndrome: A Review of Potential Mechanisms." *Cannabis and Cannabinoid Research* (2020). https://doi.org/10.1089/can.2019.0059.

DeVuono, Marieka V., Olivia La Caprara, Gavin N Petrie, Cheryl L. Limebeer, Erin M. Rock, Matthew N. Hill, and Linda A. Parker. "Cannabidiol Interferes with Establishment of $\Delta 9$—Tetrahydrocannabinol-Induced Nausea Through a 5-Ht 1a mechanism." *Cannabis and Cannabinoid Research* 7, no. 1 (2022): 58–64. http://doi.org/10.1089/can.2020.0083.

Duran, M., E. Perez, S. Abanades, X. Vidal, C. Saura, M. Majem, E. Arriola, et al. "Preliminary Efficacy and Safety of an Oromucosal Standardized Cannabis Extract in Chemotherapy-Induced Nausea and Vomiting." *British Journal of Clinical Pharmacology* 70, no. 5 (November 2010): 656–663. https://doi.org/10.1111/j.1365-2125.2010.03743.x.

ElSohly, M. A., S. Chandra, M. Radwan, C. G. Majumdar, and J. C. Church. "A Comprehensive Review of Cannabis Potency in the United States in the Last Decade." *Biological Psychiatry: Cognitive Neuroscience and Neuroimaging* 6, no. 6 (June 2021): 603–606. https://doi.org/10.1016/j.bpsc.2020.12.016.

ElSohly, M. A., Z. Mehmedic, S. Foster, C. Gon, S. Chandra, and J. C. Church. "Changes in Cannabis Potency over the Last 2 Decades (1995–2014): Analysis of Current Data in the United States." *Biological Psychiatry* 79, no. 7 (April 1, 2016): 613–619. https://doi.org/10.1016/j.biopsych.2016.01.004.

Englund, A., P. D. Morrison, J. Nottage, D. Hague, F. Kane, S. Bonaccorso, J. M. Stone, et al. "Cannabidiol Inhibits THC-Elicited Paranoid Symptoms and Hippocampal-Dependent Memory Impairment." *Journal of Psychopharmacology (Oxford, England)* 27, no. 1 (2013): 19–27. https://doi.org/10.1177/026988 1112460109; 10.1177/0269881112460109.

Fernández-Ruiz, J., E. de Lago, M. Gómez-Ruiz, C. García, O Sagredo, and M. García-Arencíbia. "Neurodegenerative Disorders Other Than Multiple Sclerosis." In *Handbook of Cannabis*, edited by Roger Pertwee, 505–525. Oxford University Press, 2014.

Fisher, Emma, R. Andrew Moore, Alexandra E. Fogarty, David P. Finn, Nanna B. Finnerup, Ian Gilron, Simon Haroutounian, et al. "Cannabinoids, Cannabis, and Cannabis-Based Medicine for Pain Management: A Systematic Review of Randomised Controlled Trials." *Pain* 162 (2021): S45–S66.

Fowler, Christopher J. "The Potential of Inhibitors of Endocannabinoid Metabolism for Drug Development: A Critical Review." *Endocannabinoids* (2015): 95–128.

Fraser, G. A. "The Use of a Synthetic Cannabinoid in the Management of Treatment-Resistant Nightmares in Posttraumatic Stress Disorder (PTSD)." *CNS Neuroscience and Therapeutics* 15, no. 1 (Winter 2009): 84–88. https://doi.org/10.1111/j.1755-5949.2008.00071.x.

Freels, Timothy G., Lydia N. Baxter-Potter, Janelle M. Lugo, Nicholas C. Glodosky, Hayden R. Wright, Samantha L. Baglot, Gavin N. Petrie, et al. "Vaporized Cannabis Extracts Have Reinforcing Properties and Support Conditioned Drug-Seeking Behavior in Rats." *Journal of Neuroscience* 40, no. 9 (2020): 1897–1908.

Fuss, Johannes, Jörg Steinle, Laura Bindila, Matthias K. Auer, Hartmut Kirchherr, Beat Lutz, and Peter Gass. "A Runner's High Depends on Cannabinoid Receptors in Mice." *Proceedings of the National Academy of Sciences USA* 112, no. 42 (2015): 13105–13108.

Gage, S. H., H. J. Jones, S. Burgess, J. Bowden, G. Davey Smith, S. Zammit, and M. R. Munafò. "Assessing Causality in Associations Between Cannabis Use and Schizophrenia Risk: A Two-Sample Mendelian Randomization Study." *Psychological Medicine* 47, no. 5 (April 2017): 971–980. https://doi.org/10.1017/s0033291716003172.

Gage, S. H., M. R. Munafò, J. MacLeod, M. Hickman, and G. D. Smith. "Cannabis and Psychosis." *Lancet Psychiatry* 2, no. 5 (May 2015): 380. https://doi.org/10.1016/s2215-0366(15)00108-x.

Gaoni, Y., and R. Mechoulam. "Isolation, Structure, and Partial Synthesis of an Active Constituent of Hashish." *Journal of the American Chemical Society* 86, no. 8 (1964): 1646–1647. https://doi.org/10.1021/ja01062a046. http://dx.doi.org/10.1021/ja01062a046.

Gaston, Tyler E., E. Martina Bebin, Gary R. Cutter, Steve B. Ampah, Yuliang Liu, Leslie P. Grayson, and Jerzy P. Szaflarski. "Drug–Drug Interactions with Cannabidiol (CBD) Appear to Have No Effect on Treatment Response in an Open-Label Expanded Access Program." *Epilepsy and Behavior* 98 (2019): 201–206.

Gates, Peter J, Lucy Albertella, and Jan Copeland. "The Effects of Cannabinoid Administration on Sleep: A Systematic Review of Human Studies." *Sleep Medicine Reviews* 18, no. 6 (2014): 477–487.

Geffrey, Alexandra L., Sarah F. Pollack, Patricia L. Bruno, and Elizabeth A. Thiele. "Drug–Drug Interaction Between Clobazam and Cannabidiol in Children with Refractory Epilepsy." *Epilepsia* 56, no. 8 (2015): 1246–1251.

Gieringer, Dale, Joseph St. Laurent, and Scott Goodrich. "Cannabis Vaporizer Combines Efficient Delivery of THC with Effective Suppression of Pyrolytic Compounds." *Journal of Cannabis Therapeutics* 4, no. 1 (2004): 7–27.

Grimison, P., A. Mersiades, A. Kirby, N. Lintzeris, R. Morton, P. Haber, I. Olver, et al. "Oral THC:CBD Cannabis Extract for Refractory Chemotherapy-Induced Nausea and Vomiting: A Randomised, Placebo-Controlled, Phase II Crossover Trial." *Annals of Oncology* 31, no. 11 (November 2020): 1553–1560. https://doi.org/10.1016/j.annonc.2020.07.020.

Guindon, J., and A. G. Hohmann. "Cannabinoid CB2 Receptors: A Therapeutic Target for the Treatment of Inflammatory and Neuropathic Pain." *British Journal of Pharmacology* 153, no. 2 (January 2008): 319–334. https://doi.org/10.1038/sj.bjp.0707531.

Gunasekera, B., K. Diederen, and S. Bhattacharyya. "Cannabinoids, Reward Processing, and Psychosis." *Psychopharmacology (Berlin)* 239, no. 5 (May 2022): 1157–1177. https://doi.org/10.1007/s00213-021-05801-2.

Haney, M. "Perspectives on Cannabis Research-Barriers and Recommendations." *JAMA Psychiatry* 77, no. 10 (October 1, 2020): 994–995. https://doi.org/10.1001/jamapsychiatry.2020.1032.

Haney, M., Z. D. Cooper, G. Bedi, S. K. Vosburg, S. D. Comer, and R. W. Foltin. "Nabilone Decreases Marijuana Withdrawal and a Laboratory Measure of Marijuana Relapse." *Neuropsychopharmacology* 38, no. 8 (July 2013): 1557–1565. https://doi.org/10.1038/npp.2013.54.

Harper, J., S. M. Malone, S. Wilson, R. H. Hunt, K. M. Thomas, and W. G. Iacono. "The Effects of Alcohol and Cannabis Use on the Cortical Thickness of Cognitive Control and Salience Brain Networks in Emerging Adulthood: A Co-Twin Control Study." *Biological Psychiatry* 89, no. 10 (May 15, 2021): 1012–1022. https://doi.org/10.1016/j.biopsych.2021.01.006.

Hartman, Rebecca L, and Marilyn A Huestis. "Cannabis Effects on Driving Skills." *Clinical Chemistry* 59, no. 3 (2013): 478–492.

Hazekamp, Arno, Mark A. Ware, Kirsten R. Muller-Vahl, Donald Abrams, and Franjo Grotenhermen. "The Medicinal Use of Cannabis and Cannabinoids—An International Cross-Sectional Survey on Administration Forms." *Journal of Psychoactive Drugs* 45, no. 3 (2013): 199–210.

Heifets, B. D., and P. E. Castillo. "Endocannabinoid Signaling and Long-Term Synaptic Plasticity." *Annual Review of Physiology* 71 (2009): 283–306. https://doi.org/10.1146/annurev.physiol.010908.163149.

Henquet, C., M. Di Forti, P. Morrison, R. Kuepper, and R. M. Murray. "Gene-Environment Interplay Between Cannabis and Psychosis." *Schizophrenia Bulletin* 34, no. 6 (November 2008): 1111–1121. https://doi.org/10.1093/schbul/sbn108.

Herkenham, Mabl, Allison B. Lynn, Mark D. Little, M. Ross Johnson, Lawrence S. Melvin, Brian R. De Costa, and Kenner C. Rice. "Cannabinoid Receptor Localization in Brain." *Proceedings of the National Academy of Sciences USA* 87, no. 5 (1990): 1932–1936.

Hill, M. N., S. Patel, P. Campolongo, J. G. Tasker, C. T. Wotjak, and J. S. Bains. "Functional Interactions Between Stress and the Endocannabinoid System: From Synaptic Signaling to Behavioral Output." *Journal of Neuroscience* 30, no. 45 (November 10, 2010): 14980–14986. https://doi.org/10.1523/jneurosci.4283-10.2010.

Hill, Matthew N., and Jeffrey G. Tasker. "Endocannabinoid Signaling, Gluco-corticoid-Mediated Negative Feedback, and Regulation of the Hypothalamic-Pituitary-Adrenal Axis." *Neuroscience* 204 (2012): 5–16.

Hindley, G., K. Beck, F. Borgan, C. E. Ginestet, R. McCutcheon, D. Kleinloog, S. Ganesh, et al. "Psychiatric Symptoms Caused by Cannabis Constituents: A Systematic Review and Meta-Analysis." *Lancet Psychiatry* 7, no. 4 (April 2020): 344–353. https://doi.org/10.1016/s2215-0366(20)30074-2.

Hindocha, C., T. P. Freeman, G. Schafer, C. Gardener, R. K. Das, C. J. Morgan, and H. V. Curran. "Acute Effects of Delta-9-Tetrahydrocannabinol, Cannabidiol and Their Combination on Facial Emotion Recognition: A Randomised, Double-Blind, Placebo-Controlled Study in Cannabis Users." *European Neuropsycho-pharmacology: The Journal of the European College of Neuropsychopharmacology* 25, no. 3 (2015): 325–334. https://doi.org/10.1016/j.euroneuro.2014.11.014; 10.1016/j.euroneuro.2014.11.014.

Hirvonen, J., R. S. Goodwin, C. T. Li, G. E. Terry, S. S. Zoghbi, C. Morse, V. W. Pike, et al. "Reversible and Regionally Selective Downregulation of Brain Can-nabinoid CB1 Receptors in Chronic Daily Cannabis Smokers." *Molecular Psy-chiatry* 17, no. 6 (June 2012): 642–649. https://doi.org/10.1038/mp.2011.82.

Hollister, L. E. "Actions of Various Marihuana Derivatives in Man." *Pharmaco-logical Reviews* 23, no. 4 (December 1971): 349–357.

Horswill, J. G., U. Bali, S. Shaaban, J. F. Keily, P. Jeevaratnam, A. J. Babbs, C. Reynet, and P. Wong Kai In. "PSNCBAM-1, a Novel Allosteric Antagonist at Cannabinoid CB1 Receptors with Hypophagic Effects in Rats." *British Journal of Pharmacology* 152, no. 5 (November 2007): 805–814. https://doi.org/10.1038/sj.bjp.0707347.

Huestis, Marilyn A., Allan Barnes, and Michael L. Smith. "Estimating the Time of Last Cannabis Use from Plasma Δ9-Tetrahydrocannabinol and 11-nor-9-Ca rboxy-Δ9-Tetrahydrocannabinol Concentrations." *Clinical Chemistry* 51, no. 12 (2005): 2289–2295.

Huestis, Marilyn A., and Michael L. Smith. "Cannabinoid Pharmacokinetics and Disposition in Alternative Matrices." In *Handbook of Cannabis*, edited by Roger Pertwee, 296–316. Oxford University Press, 2014.

Hughes, J., T. W. Smith, H.W. Kosterlitz, Linda A. Fothergill, B. A. Morgan, and H. R. Morris. "Identification of Two Related Pentapeptides from the Brain with Potent Opiate Agonist Activity." *Nature* 258, no. 5536 (1975): 577–579.

Hurd, Y. L., M. Yoon, A. F. Manini, S. Hernandez, R. Olmedo, M. Ostman, and D. Jutras-Aswad. "Early Phase in the Development of Cannabidiol as a Treatment for Addiction: Opioid Relapse Takes Initial Center Stage." *Neurotherapeutics* 12, no. 4 (October 2015): 807–815. https://doi.org/10.1007/s13311-015-0373-7.

Hurd, Y. L., S. Spriggs, J. Alishayev, G. Winkel, K. Gurgov, C. Kudrich, A. M. Oprescu, and E. Salsitz. "Cannabidiol for the Reduction of Cue-Induced Craving and Anxiety in Drug-Abstinent Individuals with Heroin Use Disorder: A Double-Blind Randomized Placebo-Controlled Trial." *American Journal of Psychiatry* 176, no. 11 (November 1, 2019): 911–922. https://doi.org/10.1176/appi.ajp.2019.18101191.

Häuser, W., P. Welsch, L. Radbruch, E. Fisher, R. F. Bell, and R. A. Moore. "Cannabis-Based Medicines and Medical Cannabis for Adults with Cancer Pain." *Cochrane Database System Rev* 6, no. 6 (June 5 2023): Cd014915. https://doi.org/10.1002/14651858.CD014915.pub2.

Ignatowska-Jankowska, B. M., G. L. Baillie, S. Kinsey, M. Crowe, S. Ghosh, R. A. Owens, I. M. Damaj, et al. "A Cannabinoid CB1 Receptor-Positive Allosteric Modulator Reduces Neuropathic Pain in the Mouse with No Psychoactive Effects." *Neuropsychopharmacology* 40, no. 13 (December 2015): 2948–2959. https://doi.org/10.1038/npp.2015.148.

Jackson, N. J., J. D. Isen, R. Khoddam, D. Irons, C. Tuvblad, W. G. Iacono, M. McGue, A. Raine, and L. A. Baker. "Impact of Adolescent Marijuana Use on Intelligence: Results from Two Longitudinal Twin Studies." *Proceedings of the National Academy of Sciences USA* 113, no. 5 (February 2, 2016): E500–E508. https://doi.org/10.1073/pnas.1516648113.

Jetly, R., A. Heber, G. Fraser, and D. Boisvert. "The Efficacy of Nabilone, a Synthetic Cannabinoid, in the Treatment of PTSD-Associated Nightmares: A Preliminary Randomized, Double-Blind, Placebo-Controlled Cross-Over Design Study." *Psychoneuroendocrinology* 51 (January 2015): 585–588. https://doi.org/10.1016/j.psyneuen.2014.11.002.

Jones, G., and R. G. Pertwee. "A Metabolic Interaction in Vivo Between Cannabidiol and $\Delta 1$-Tetrahydrocannabinol." *British Journal of Pharmacology* 45, no. 2 (1972): 375–377.

Jung, K. M., J. R. Clapper, J. Fu, G. D'Agostino, A. Guijarro, D. Thongkham, A. Avanesian, et al. "2-Arachidonoylglycerol Signaling in Forebrain Regulates Systemic Energy Metabolism." *Cell Metabolism* 15, no. 3 (March 7, 2012): 299–310. https://doi.org/10.1016/j.cmet.2012.01.021.

Justinova, Z., R. A. Mangieri, M. Bortolato, S. I. Chefer, A. G. Mukhin, J. R. Clapper, A. R. King, et al. "Fatty Acid Amide Hydrolase Inhibition Heightens Anandamide Signaling Without Producing Reinforcing Effects in Primates." *Biological Psychiatry* 64, no. 11 (December 1 2008): 930–937. https://doi.org /10.1016/j.biopsych.2008.08.008.

King, Kirsten M., Alyssa M. Myers, Ariele J. Soroka-Monzo, Ronald F. Tuma, Ronald J. Tallarida, Ellen A. Walker, and Sara Jane Ward. "Single and Combined Effects of Δ9-Tetrahydrocannabinol and Cannabidiol in a Mouse Model of Chemotherapy-Induced Neuropathic Pain." *British Journal of Pharmacology* 174, no. 17 (2017): 2832–2841.

Laprairie, R. B., A. M. Bagher, M. E. Kelly, and E. M. Denovan-Wright. "Cannabidiol Is a Negative Allosteric Modulator of the Cannabinoid CB1 Receptor." *British Journal of Pharmacology* 172, no. 20 (October 2015): 4790–4805. https://doi.org/10.1111/bph.13250.

Large, M., S. Sharma, M. T. Compton, T. Slade, and O. Nielssen. "Cannabis Use and Earlier Onset of Psychosis: A Systematic Meta-Analysis." *Archives of General Psychiatry* 68, no. 6 (June 2011): 555–561. https://doi.org/10.1001 /archgenpsychiatry.2011.5.

Leweke, F. M., D. Piomelli, F. Pahlisch, D. Muhl, C. W. Gerth, C. Hoyer, J. Klosterkotter, M. Hellmich, and D. Koethe. "Cannabidiol Enhances Anandamide Signaling and Alleviates Psychotic Symptoms of Schizophrenia." *Translational Psychiatry* 2 (March 20, 2012): e94. https://doi.org/10.1038/tp .2012.15.

Limebeer, C. L., E. M. Rock, K. A. Sharkey, and L. A. Parker. "Nausea-Induced 5-HT Release in the Interoceptive Insular Cortex and Regulation by Monoacylglycerol Lipase (MAGL) Inhibition and Cannabidiol." *eNeuro* 5, no. 4 (July–August 2018). https://doi.org/10.1523/eneuro.0256-18.2018.

Lötsch, J., I. Weyer-Menkhoff, and I. Tegeder. "Current Evidence of Cannabinoid-Based Analgesia Obtained in Preclinical and Human Experimental Settings." *European Journal of Pain* 22, no. 3 (March 2018): 471–484. https://doi .org/10.1002/ejp.1148.

Maccarrone, M., V. Di Marzo, J. Gertsch, U. Grether, A. C. Howlett, T. Hua, A. Makriyannis, et al. "Goods and Bads of the Endocannabinoid System as a Therapeutic Target: Lessons Learned After 30 Years." *Pharmacological Review* 75, no. 5 (September 2023): 885–958. https://doi.org/10.1124/pharmrev.122.000600.

Manini, A. F., G. Yiannoulos, M. M. Bergamaschi, S. Hernandez, R. Olmedo, A. J. Barnes, G. Winkel, et al. "Safety and Pharmacokinetics of Oral Cannabidiol When Administered Concomitantly with Intravenous Fentanyl in Humans." *Journal of Addiction Medicine* 9, no. 3 (May–June 2015): 204–210. https://doi .org/10.1097/adm.0000000000000118.

Marsicano, G., C. T. Wotjak, S. C. Azad, T. Bisogno, G. Rammes, M. G. Cascio, H. Hermann, et al. "The Endogenous Cannabinoid System Controls Extinction of Aversive Memories." *Nature* 418, no. 6897 (August 1, 2002): 530–534. https:// doi.org/10.1038/nature00839.

Marsicano, G., and P. Lafenêtre. "Roles of the Endocannabinoid System in Learning and Memory." *Current Topics in Behavioral Neuroscience* 1 (2009): 201– 230. https://doi.org/10.1007/978-3-540-88955-7_8.

Martin, B. R., D. R. Compton, B. F. Thomas, W. R. Prescott, P. J. Little, R. K. Razdan, M. R. Johnson, et al. "Behavioral, Biochemical, and Molecular Modeling Evaluations of Cannabinoid Analogs." *Pharmacology, Biochemistry, and Behavior* 40, no. 3 (1991): 471–478.

McGuire, P., P. Robson, W. J. Cubala, D. Vasile, P. D. Morrison, R. Barron, A. Taylor, and S. Wright. "Cannabidiol (CBD) as an Adjunctive Therapy in Schizophrenia: A Multicenter Randomized Controlled Trial." *American Journal of Psychiatry* 175, no. 3 (March 1, 2018): 225–231. https://doi.org/10.1176/appi .ajp.2017.17030325.

Mead, Alice P. "International Control of Cannabis." In *Handbook of Cannabis*, edited by Roger Pertwee, 44–64. Oxford University Press, 2014.

Mechoulam, R., and E. A. Carlini. "Toward Drugs Derived from Cannabis." *Die Naturwissenschaften* 65, no. 4 (1978): 174–179.

Mechoulam, R., L. O. Hanuš, R. Pertwee, and A. C. Howlett. "Early Phytocannabinoid Chemistry to Endocannabinoids and Beyond." *Nature Reviews Neuroscience* 15, no. 11 (2014): 757–764..

Mechoulam, R., S. Ben-Shabat, L. Hanus, M. Ligumsky, N. E. Kaminski, A. R. Schatz, A. Gopher, et al. "Identification of an Endogenous 2-Monoglyceride, Present in Canine Gut, That Binds to Cannabinoid Receptors." *Biochemical Pharmacology* 50, no. 1 (1995): 83–90.

Mechoulam, R., and Y. Shvo. "Hashish. I. The Structure of Cannabidiol." *Tetrahedron* 19, no. 12 (December 1963): 2073–2078. https://doi.org/10.1016 /0040-4020(63)85022-x.

Meier, M. H., A. Caspi, A. Ambler, H. Harrington, R. Houts, R. S. Keefe, K. McDonald, et al. "Persistent Cannabis Users Show Neuropsychological Decline from Childhood to Midlife." *Proceedings of the National Academy of Sciences USA* 109, no. 40 (October 2, 2012): E2657– E2664. https://doi.org/10.1073/pnas.1206820109.

Meiri, E., H. Jhangiani, J. J. Vredenburgh, L. M. Barbato, F. J. Carter, H. M. Yang, and V. Baranowski. "Efficacy of Dronabinol Alone and in Combination with Ondansetron Versus Ondansetron Alone for Delayed Chemotherapy-Induced Nausea and Vomiting." *Current Medical Research and Opinion* 23, no. 3 (March 2007): 533–543. https://doi.org/10.1185/030079907x167525.

Miranda, Alannah, Elizabeth Peek, Sonia Ancoli-Israel, Jared W. Young, William Perry, and Arpi Minassian. "The Role of Cannabis and the Endocannabinoid System in Sleep Regulation and Cognition: A Review of Human and Animal Studies." *Behavioral Sleep Medicine* 22, no. 2 (2024): 217–233.

Mittleman, Murray A., Rebecca A. Lewis, Malcolm Maclure, Jane B. Sherwood, and James E. Muller. "Triggering Myocardial Infarction by Marijuana." *Circulation* 103, no. 23 (2001): 2805–2809.

Mokrysz, C., R. Landy, S. H. Gage, M. R. Munafò, J. P. Roiser, and H. V. Curran. "Are IQ and Educational Outcomes in Teenagers Related to Their Cannabis Use? A Prospective Cohort Study." *Journal of Psychopharmacology* 30, no. 2 (February 2016): 159–168. https://doi.org/10.1177/0269881115622241.

Monteleone, P., I. Matias, V. Martiadis, L. De Petrocellis, M. Maj, and V. Di Marzo. "Blood Levels of the Endocannabinoid Anandamide Are Increased in Anorexia Nervosa and in Binge-Eating Disorder, but Not in Bulimia Nervosa." *Neuropsychopharmacology* 30, no. 6 (June 2005): 1216–1221. https://doi.org/10.1038/sj.npp.1300695.

Moore, T. H., S. Zammit, A. Lingford-Hughes, T. R. Barnes, P. B. Jones, M. Burke, and G. Lewis. "Cannabis Use and Risk of Psychotic or Affective Mental Health Outcomes: A Systematic Review." *Lancet* 370, no. 9584 (July 28, 2007): 319–328. https://doi.org/10.1016/s0140-6736(07)61162-3.

Moreau, J. J. *Hashish and Mental Illness*, translated from French by Gordon J. Barnett. New York: Raven, 1973 (1845).

Morales, P., D. P. Hurst, and P. H. Reggio. "Molecular Targets of the Phytocannabinoids: A Complex Picture." *Progress in the Chemistry of Organic Natural Products* 103 (2017): 103–131. https://doi.org/10.1007/978-3-319-45541-9_4.

Morena, M., S. Patel, J. S. Bains, and M. N. Hill. "Neurobiological Interactions Between Stress and the Endocannabinoid System." *Neuropsychopharmacology* 41, no. 1 (January 2016): 80–102. https://doi.org/10.1038/npp.2015.166.

Morgan, C. J. A., T. P. Freeman, C. Hindocha, G. Schafer, C. Gardner, and H. V. Curran. "Individual and Combined Effects of Acute Delta-9-Tetrahydrocannabinol and Cannabidiol on Psychotomimetic Symptoms and Memory Function." *Translational Psychiatry* 8, no. 1 (September 5, 2018): 181. https://doi.org/10.1038/s41398-018-0191-x.

Munro, S., K. L. Thomas, and M. Abu-Shaar. "Molecular Characterization of a Peripheral Receptor for Cannabinoids." *Nature* 365, no. 6441 (September 2, 1993): 61–65. https://doi.org/10.1038/365061a0.

Napadow, V., J. D. Sheehan, J. Kim, L. T. Lacount, K. Park, T. J. Kaptchuk, B. R. Rosen, and B. Kuo. "The Brain Circuitry Underlying the Temporal Evolution of Nausea in Humans." *Cerebral Cortex* 23, no. 4 (April 2013): 806–813. https://doi.org/10.1093/cercor/bhs073.

O'Shaughnessy, W. B. "On the Preparations of Indian Hemp, or Gunja (*Cannabis Indica*); Their Effects on the Animal System in Health, and Their Utility in the Treatment of Tentanus and Other Convulsive Diseases." *Transactions of the Medical and Physical Society of Bengal* (1839): 421–461.

Pacher, P., and R. Mechoulam. "Is Lipid Signaling Through Cannabinoid 2 Receptors Part of a Protective System?" *Progress in Lipid Research* 50, no. 2 (2011): 193–211. https://doi.org/10.1016/j.plipres.2011.01.001; 10.1016/j.plipres.2011.01.001.

Paes-Colli, Yolanda, Andrey F. L. Aguiar, Alinny Rosendo Isaac, Bruna K. Ferreira, Raquel Maria P. Campos, Priscila Martins Pinheiro Trindade, Ricardo Augusto de Melo Reis, and Luzia S Sampaio. "Phytocannabinoids and Cannabis-Based Products as Alternative Pharmacotherapy in Neurodegenerative Diseases: From Hypothesis to Clinical Practice." *Frontiers in Cellular Neuroscience* 16 (2022): 917164.

Parker, Linda A., Erin M. Rock, and Raphael Mechoulam. *CBD: What Does the Science Say?* MIT Press, 2022.

Parsons, Loren H, and Yasmin L Hurd. "Endocannabinoid Signalling in Reward and Addiction." *Nature Reviews Neuroscience* 16, no. 10 (2015): 579–594.

Pasman, J. A., K. J. H. Verweij, Z. Gerring, S. Stringer, S. Sanchez-Roige, J. L. Treur, A. Abdellaoui, et al. "GWAS of Lifetime Cannabis Use Reveals New Risk

Loci, Genetic Overlap with Psychiatric Traits, and a Causal Influence of Schizophrenia." *Nature Neuroscience* 21, no. 9 (September 2018): 1161–1170. https://doi.org/10.1038/s41593-018-0206-1.

Patel, Sachin, Matthew N. Hill, and Cecilia J. Hillard. "Effects of Phytocannabinoids on Anxiety, Mood, and the Endocrine System." In *Handbook of Cannabis*, edited by Roger Pertwee, 189–207. Oxford University Press, 2014.

Patrician, Alexander, Maja Versic-Bratincevic, Tanja Mijacika, Ivana Banic, Mario Marendic, Davorka Sutlović, Željko Dujić, and Philip N Ainslie. "Examination of a New Delivery Approach for Oral Cannabidiol in Healthy Subjects: A Randomized, Double-Blinded, Placebo-Controlled Pharmacokinetics Study." *Advances in Therapy* 36 (2019): 3196–3210.

Paul, S. E., A. S. Hatoum, J. D. Fine, E. C. Johnson, I. Hansen, N. R. Karcher, A. L. Moreau, et al. "Associations Between Prenatal Cannabis Exposure and Childhood Outcomes: Results from the ABCD Study." *JAMA Psychiatry* 78, no. 1 (January 1, 2021): 64–76. https://doi.org/10.1001/jamapsychiatry.2020.2902.

Pertwee, Roger G. "Inverse Agonism and Neutral Antagonism at Cannabinoid CB1 Receptors." *Life Sciences* 76, no. 12 (2005): 1307–1324.

Piomelli, Daniele (moderator), and Margaret Haney, Alan J. Budney, and Pier Vincenzo Piazza (participants). "Legal or Illegal, Cannabis Is Still Addictive." *Cannabis and Cannabinoid Research* 1, no. 1 (2016): 47–53.

Pi-Sunyer, F. X., L. J. Aronne, H. M. Heshmati, J. Devin, J. Rosenstock, and RIO-North America Study Group. "Effect of Rimonabant, a Cannabinoid-1 Receptor Blocker, on Weight and Cardiometabolic Risk Factors in Overweight or Obese Patients: RIO-North America: A Randomized Controlled Trial." *JAMA* 295, no. 7 (2006): 761–775. https://doi.org/10.1001/jama.295.7.761.

Pisanti, S., and M. Bifulco. "Medical Cannabis: A Plurimillennial History of an Evergreen." *Journal of Cellular Physiology* 234, no. 6 (June 2019): 8342–8351. https://doi.org/10.1002/jcp.27725.

Portenoy, R. K., E. D. Ganae-Motan, S. Allende, R. Yanagihara, L. Shaiova, S. Weinstein, R. McQuade, S. Wright, and M. T. Fallon. "Nabiximols for Opioid-Treated Cancer Patients with Poorly-Controlled Chronic Pain: A Randomized, Placebo-Controlled, Graded-Dose Trial." *Journal of Pain* 13, no. 5 (May 2012): 438–449. https://doi.org/10.1016/j.jpain.2012.01.003.

Power, R. A., K. J. Verweij, M. Zuhair, G. W. Montgomery, A. K. Henders, A. C. Heath, P. A. Madden, et al. "Genetic Predisposition to Schizophrenia Associated with Increased Use of Cannabis." *Molecular Psychiatry* 19, no. 11 (November 2014): 1201–1204. https://doi.org/10.1038/mp.2014.51.

Pryce, G., and D. Baker. "Cannabis and Multiple Sclerosis." In *Handbook of Cannabis*, edited by Roger Pertwee, 487–504. Oxford University Press, 2014.

Ranum, Rylea M., Mary O. Whipple, Ivana Croghan, Brent Bauer, Loren L. Toussaint, and Ann Vincent. "Use of Cannabidiol in the Management of Insomnia: A Systematic Review." *Cannabis and Cannabinoid Research* 8, no. 2 (2023): 213–229.

Ren, Y., J. Whittard, A. Higuera-Matas, C. V. Morris, and Y. L. Hurd. "Cannabidiol, a Nonpsychotropic Component of Cannabis, Inhibits Cue-Induced Heroin Seeking and Normalizes Discrete Mesolimbic Neuronal Disturbances." *Journal of Neuroscience: The Official Journal of the Society for Neuroscience* 29, no. 47 (2009): 14764–14769. https://doi.org/10.1523/JNEUROSCI.4291-09.2009.

Renard, J., C. Norris, W. Rushlow, and S. R. Laviolette. "Neuronal and Molecular Effects of Cannabidiol on the Mesolimbic Dopamine System: Implications for Novel Schizophrenia Treatments." *Neuroscience and Biobehavioral Review* 75 (April 2017): 157–165. https://doi.org/10.1016/j.neubiorev.2017.02.006.

Rock, E. M., C. L. Limebeer, R. G. Pertwee, R. Mechoulam, and L. A. Parker. "Therapeutic Potential of Cannabidiol, Cannabidiolic Acid, and Cannabidiolic Acid Methyl Ester as Treatments for Nausea and Vomiting." *Cannabis and Cannabinoid Research* 6, no. 4 (August 2021): 266–274. https://doi.org/10.1089/can.2021.0041.

Rock, E. M., D. Bolognini, C. L. Limebeer, M. G. Cascio, S. Anavi-Goffer, P. J. Fletcher, R. Mechoulam, R. G. Pertwee, and L. A. Parker. "Cannabidiol, a Non-Psychotropic Component of Cannabis, Attenuates Vomiting and Nausea-Like Behaviour Via Indirect Agonism of 5-Ht(1a) Somatodendritic Autoreceptors in the Dorsal Raphe Nucleus." *British Journal of Pharmacology* 165, no. 8 (2012): 2620–2634. https://doi.org/10.1111/j.1476-5381.2011.01621.x; 10.1111/j.1476-5381.2011.01621.x.

Rock, E. M., and L. A. Parker. "Cannabinoids as Potential Treatment for Chemotherapy-Induced Nausea and Vomiting." *Frontiers in Pharmacology* 7 (2016): 221. https://doi.org/10.3389/fphar.2016.00221.

———. "Synergy Between Cannabidiol, Cannabidiolic Acid, and Delta(9)-Tetrahydrocannabinol in the Regulation of Emesis in the Suncus Murinus

(House Musk Shrew)." *Behavioral Neuroscience* 129, no. 3 (2015): 368–370. https://doi.org/10.1037/bne0000057; 10.1037/bne0000057.

Rock, E. M., M. T. Sullivan, S. Pravato, M. Pratt, C. L. Limebeer, and L. A. Parker. "Effect of Combined Doses of Δ(9)-Tetrahydrocannabinol and Cannabidiol or Tetrahydrocannabinolic Acid and Cannabidiolic Acid on Acute Nausea in Male Sprague-Dawley Rats." *Psychopharmacology (Berlin)* 237, no. 3 (March 2020): 901–914. https://doi.org/10.1007/s00213-019-05428-4.

Russo, E. B. "Cannabis and Epilepsy: An Ancient Treatment Returns to the Fore." *Epilepsy and Behavior* 70, Pt. B (May 2017): 292–297. https://doi.org /10.1016/j.yebeh.2016.09.040.

Russo, E., and G. W. Guy. "A Tale of Two Cannabinoids: The Therapeutic Rationale for Combining Tetrahydrocannabinol and Cannabidiol." *Medical Hypotheses* 66, no. 2 (2006): 234–246. https://doi.org/10.1016/j.mehy.2005.08.026.

Russo, Ethan B. "Taming THC: Potential Cannabis Synergy and Phytocannabinoid-Terpenoid Entourage Effects." *British Journal of Pharmacology* 163, no. 7 (2011): 1344–1364.

Sapolsky, R. M. *Behave: The Biology of Humans at Our Best and Worst*. Penguin Publishing Group, 2017.

Scheen, A. J., N. Finer, P. Hollander, M. D. Jensen, and L. F. Van Gaal. "Efficacy and Tolerability of Rimonabant in Overweight or Obese Patients with Type 2 Diabetes: A Randomised Controlled Study." *Lancet* 368, no. 9548 (November 11, 2006): 1660–1672. https://doi.org/10.1016/s0140-6736(06)69571-8.

Schmidt, M. E., M. R. Liebowitz, M. B. Stein, J. Grunfeld, I. Van Hove, W. K. Simmons, P. Van Der Ark, et al. "The Effects of Inhibition of Fatty Acid Amide Hydrolase (FAAH) by Jnj-42165279 in Social Anxiety Disorder: A Double-Blind, Randomized, Placebo-Controlled Proof-of-Concept Study." *Neuropsychopharmacology* 46, no. 5 (April 2021): 1004–1010. https://doi.org/10.1038 /s41386-020-00888-1.

Schoedel, Kerri Alexandra, Nancy Chen, Annie Hilliard, Linda White, Colin Stott, Ethan Russo, Stephen Wright, et al. "A Randomized, Double-Blind, Placebo-Controlled, Crossover Study to Evaluate the Subjective Abuse Potential and Cognitive Effects of Nabiximols Oromucosal Spray in Subjects with a History of Recreational Cannabis Use." *Human Psychopharmacology: Clinical and Experimental* 26, no. 3 (2011): 224–236.

Schreiner, A. M., and M. E. Dunn. "Residual Effects of Cannabis Use on Neurocognitive Performance After Prolonged Abstinence: A Meta-Analysis." *Experimental and Clinical Psychopharmacology* 20, no. 5 (October 2012): 420–429. https://doi.org/10.1037/a0029117.

Scicluna, Rhianne L., Bianca B. Wilson, Samuel H. Thelaus, Jonathon C. Arnold, Iain S. McGregor, and Michael T. Bowen. "Cannabidiol Reduced the Severity of Gastrointestinal Symptoms of Opioid Withdrawal in Male and Female Mice." *Cannabis and Cannabinoid Research* 9, no. 2 (2024): 547–560.

Sclocco, R., J. Kim, R. G. Garcia, J. D. Sheehan, F. Beissner, A. M. Bianchi, S. Cerutti, et al. "Brain Circuitry Supporting Multi-Organ Autonomic Outflow in Response to Nausea." *Cerebral Cortex* 26, no. 2 (February 2016): 485–497. https://doi.org/10.1093/cercor/bhu172.

Silveira, Mason M., Jonathon C. Arnold, Steven R. Laviolette, Cecilia J. Hillard, Marta Celorrio, Maria S Aymerich, and Wendy K. Adams. "Seeing Through the Smoke: Human and Animal Studies of Cannabis Use and Endocannabinoid Signalling in Corticolimbic Networks." *Neuroscience and Biobehavioral Reviews* 76 (2017): 380–395.

Sink, K. S., P. J. McLaughlin, J. A. Wood, C. Brown, P. Fan, V. K. Vemuri, Y. Peng, et al. "The Novel Cannabinoid CB1 Receptor Neutral Antagonist Am4113 Suppresses Food Intake and Food-Reinforced Behavior but Does Not Induce Signs of Nausea in Rats." *Neuropsychopharmacology* 33, no. 4 (2008): 946–955. https://doi.org/10.1038/sj.npp.1301476.

Soler-Cedeño, Omar, Hannah Alton, Guo-Hua Bi, Emily Linz, Lipin Ji, Alexandros Makriyannis, and Zheng-Xiong Xi. "Am6527, a Neutral CB1 Receptor Antagonist, Suppresses Opioid Taking and Seeking, as Well as Cocaine Seeking in Rodents without Aversive Effects." *Neuropsychopharmacology* (2024): 1–11.

Solowij, N., S. J. Broyd, C. Beale, J. A. Prick, L. M. Greenwood, H. van Hell, C. Suo, et al. "Therapeutic Effects of Prolonged Cannabidiol Treatment on Psychological Symptoms and Cognitive Function in Regular Cannabis Users: A Pragmatic Open-Label Clinical Trial." *Cannabis and Cannabinoid Researchs* 3, no. 1 (2018): 21–34. https://doi.org/10.1089/can.2017.0043.

Soltesz, I., B. E. Alger, M. Kano, S. H. Lee, D. M. Lovinger, T. Ohno-Shosaku, and M. Watanabe. "Weeding Out Bad Waves: Towards Selective Cannabinoid Circuit Control in Epilepsy."*Nature Reviews Neuroscience* 16, no. 5 (May 2015): 264–277. https://doi.org/10.1038/nrn3937.

Spanagel, Rainer, and Ainhoa Bilbao. "Approved Cannabinoids for Medical Purposes—Comparative Systematic Review and Meta-Analysis for Sleep and Appetite." *Neuropharmacology* 196 (2021): 108680.

Sticht, M. A., C. L. Limebeer, B. R. Rafla, R. A. Abdullah, J. L. Poklis, W. Ho, M. J. Niphakis, et al. "Endocannabinoid Regulation of Nausea Is Mediated by 2-Arachidonoylglycerol (2-AG) in the Rat Visceral Insular Cortex." *Neuropharmacology* 102 (March 2016): 92–102. https://doi.org/10.1016/j.neuropharm.2015.10.039.

Stockings, Emily, Gabrielle Campbell, Wayne D. Hall, Suzanne Nielsen, Dino Zagic, Rakin Rahman, Bridin Murnion, et al. "Cannabis and Cannabinoids for the Treatment of People with Chronic Noncancer Pain Conditions: A Systematic Review and Meta-Analysis of Controlled and Observational Studies." *Pain* 159, no. 10 (2018): 1932–1954.

Sugiura, Takayuki, Sachiko Kondo, Akihiro Sukagawa, Shinji Nakane, Akira Shinoda, Kiyoko \sItoh, Atsushi Yamashita, and Keizo Waku. "2-Arachidonoylgylcerol: A Possible Endogenous Cannabinoid Receptor Ligand in Brain." *Biochemical and Biophysical Research Communications* 215, no. 1 (1995): 89–97.

Tam, J., R. Cinar, J. Liu, G. Godlewski, D. Wesley, T. Jourdan, G. Szanda, et al. "Peripheral Cannabinoid-1 Receptor Inverse Agonism Reduces Obesity by Reversing Leptin Resistance." *Cell Metabolism* 16, no. 2 (August 8, 2012): 167–179. https://doi.org/10.1016/j.cmet.2012.07.002.

Thiele, E. A., E. D. Marsh, J. A. French, M. Mazurkiewicz-Beldzinska, S. R. Benbadis, C. Joshi, P. D. Lyons, et al. "Cannabidiol in Patients with Seizures Associated with Lennox-Gastaut Syndrome (GWPCARE4): A Randomised, Double-Blind, Placebo-Controlled Phase 3 Trial."*Lancet* 391, no. 10125 (March 17, 2018): 1085–1096. https://doi.org/10.1016/s0140-6736(18)30136-3.

Thomas, A., L. A. Stevenson, K. N. Wease, M. R. Price, G. Baillie, R. A. Ross, and R. G. Pertwee. "Evidence That the Plant Cannabinoid Delta9-Tetrahydrocannabivarin Is a Cannabinoid CB1 and CB2 Receptor Antagonist." *British Journal of Pharmacology* 146, no. 7 (2005): 917–926. https://doi.org/10.1038/sj.bjp.0706414.

Urbi, Berzenn, Joel Corbett, Ian Hughes, Maame Amma Owusu, Sarah Thorning, Simon A. Broadley, Arman Sabet, and Saman Heshmat. "Effects of Cannabis in Parkinson's Disease: A Systematic Review and Meta-Analysis." *Journal of Parkinson's Disease* 12, no. 2 (2022): 495–508.

Van Gaal, L. F., A. M. Rissanen, A. J. Scheen, O. Ziegler, and S. Rössner. "Effects of the Cannabinoid-1 Receptor Blocker Rimonabant on Weight Reduction and Cardiovascular Risk Factors in Overweight Patients: 1-Year Experience from the RIO-Europe Study." *Lancet* 365, no. 9468 (April 16–22, 2005): 1389–1397. https://doi.org/10.1016/s0140-6736(05)66374-x.

Varvel, S. A., J. L. Wiley, R. Yang, D. T. Bridgen, K. Long, A. H. Lichtman, and B. R. Martin. "Interactions Between THC and Cannabidiol in Mouse Models of Cannabinoid Activity." *Psychopharmacology (Berlin)* 186, no. 2 (June 2006): 226–234. https://doi.org/10.1007/s00213-006-0356-9.

Vigil, J. M., S. S. Stith, I. M. Adams, and A. P. Reeve. "Associations Between Medical Cannabis and Prescription Opioid Use in Chronic Pain Patients: A Preliminary Cohort Study." *PLoS One* 12, no. 11 (2017): e0187795. https://doi.org/10.1371/journal.pone.0187795.

Ward, S. J., S. D. McAllister, R. Kawamura, R. Murase, H. Neelakantan, and E. A. Walker. "Cannabidiol Inhibits Paclitaxel-Induced Neuropathic Pain Through 5-Ht(1a) Receptors without Diminishing Nervous System Function or Chemotherapy Efficacy." *British Journal of Pharmacology* 171, no. 3 (2014): 636–645. https://doi.org/10.1111/bph.12439; 10.1111/bph.12439.

Weiland, B. J., R. E. Thayer, B. E. Depue, A. Sabbineni, A. D. Bryan, and K. E. Hutchison. "Daily Marijuana Use Is Not Associated with Brain Morphometric Measures in Adolescents or Adults." *Journal of Neuroscience* 35, no. 4 (January 28, 2015): 1505–1512. https://doi.org/10.1523/jneurosci.2946-14.2015.

Whiting, P. F., R. F. Wolff, S. Deshpande, M. Di Nisio, S. Duffy, A. V. Hernandez, J. C. Keurentjes, et al. "Cannabinoids for Medical Use: A Systematic Review and Meta-Analysis." *JAMA* 313, no. 24 (June 23–30, 2015): 2456–2473. https://doi.org/10.1001/jama.2015.6358.

Wilkinson, S. T., R. Radhakrishnan, and D. C. D'Souza. "Impact of Cannabis Use on the Development of Psychotic Disorders." *Current Addiction Reports* 1, no. 2 (June 1, 2014): 115–128. https://doi.org/10.1007/s40429-014-0018-7.

Williams, C. M., N. A. Jones, and B. J. Whalley. "Cannabis and Epilepsy." In *Handbook of Cannabis*, edited by Roger Pertwee, 547–563. Oxford University Press, 2014.

Wilsey, B., T. Marcotte, R. Deutsch, B. Gouaux, S. Sakai, and H. Donaghe. "Low-Dose Vaporized Cannabis Significantly Improves Neuropathic Pain." *Journal of Pain* 14, no. 2 (February 2013): 136–148. https://doi.org/10.1016/j.jpain .2012.10.009.

Winhusen, Theresa, Jeff Theobald, David C Kaelber, and Daniel Lewis. "Regular Cannabis Use, with and Without Tobacco Co-Use, Is Associated with Respiratory Disease." *Drug and Alcohol Dependence* 204 (2019): 107557.

Wood, T. Barlow, W. T. Newton Spivey, and Thomas Hill Easterfield. "Xl.— Charas. The Resin of Indian Hemp." *Journal of the Chemical Society, Transactions* 69 (1896): 539–546.

Zajicek, J., S. Ball, D. Wright, J. Vickery, A. Nunn, D. Miller, M. G. Cano, et al. "Effect of Dronabinol on Progression in Progressive Multiple Sclerosis (Cupid): A Randomised, Placebo-Controlled Trial." *Lancet Neurology* 12, no. 9 (September 2013): 857–865. https://doi.org/10.1016/s1474-4422(13)70159-5.

Zamarripa, C. Austin, Tory R. Spindle, Renuka Surujunarain, Elise M. Weerts, Sumit Bansal, Jashvant D. Unadkat, Mary F. Paine, and Ryan Vandrey. "Assessment of Orally Administered Δ9-Tetrahydrocannabinol When Coadministered with Cannabidiol on Δ9-Tetrahydrocannabinol Pharmacokinetics and Pharmacodynamics in Healthy Adults: A Randomized Clinical Trial." *JAMA Network Open* 6, no. 2 (2023): e2254752-e52.

Zammit, S., M. J. Owen, J. Evans, J. Heron, and G. Lewis. "Cannabis, COMT and Psychotic Experiences."*British Journal of Psychiatry* 199, no. 5 (November 2011): 380–385. https://doi.org/10.1192/bjp.bp.111.091421.

Zuardi, A. W., I. Shirakawa, E. Finkelfarb, and I. G. Karniol. "Action of Cannabidiol on the Anxiety and Other Effects Produced by Delta 9-THC in Normal Subjects." *Psychopharmacology* 76, no. 3 (1982): 245–250.

Zuardi, A. W., J. A. Crippa, J. E. Hallak, J. P. Pinto, M. H. Chagas, G. G. Rodrigues, S. M. Dursun, and V. Tumas. "Cannabidiol for the Treatment of Psychosis in Parkinson's Disease." *Journal of Psychopharmacology* 23, no. 8 (November 2009): 979–983. https://doi.org/10.1177/0269881108096519.

FURTHER READING

Books

Iversen, L. L. *The Science of Marijuana*. Oxford University Press, 2008.

Mechoulam, R., ed. *Cannabinoids as Therapeutics*. Springer S, 2006.

Parker, L. A. *Cannabinoids and the Brain*. MIT Press, 2017.

Parker, L. A., and E. M. Rock, R. Mechoulam. *CBD: What Does the Science Say?* MIT Press, 2022.

Pertwee, R., ed. *Handbook of Cannabis*. Oxford University Press, 2014.

Journal Article Reviews

Bilbao, A., and R. Spanagel. "Medical Cannabinoids: A Pharmacology-Based Systematic Review and Meta-Analysis for All Relevant Medical Indications." *BMC Medicine* 20 (2022): 259-

Broyd, S. J., H. H. van Hell, C. Beale, M. Yücel, and N. Solowij. "Acute and Chronic Effects of Cannabinoids on Human Cognition-a Systematic Review." *Biological Psychiatry* 79 (2016): 557–567.

Curran, H. V., T. P. Freeman, C. Mokrysz, D. A. Lewis, C. J. Morgan, and L. H. Parsons. "Keep Off the Grass? Cannabis, Cognition and Addiction." *Nature Reviews Neuroscience* 17 (2016): 293–306.

Maccarrone, M., V. Di Marzo, J. Gertsch, U. Grether, A. C. Howlett, T. Hua, A. Makriyannis, et al. "Goods and Bads of the Endocannabinoid System as a Therapeutic Target: Lessons Learned after 30 Years." *Pharmacological Review* 75 (2023): 885–958.

Mechoulam R., and L. A. Parker. "The Endocannabinoid System and the Brain." *Annual Review of Psychology* 64 (2013): 21–47.

Mechoulam, R., L. O. Hanuš, R. Pertwee, and A. C. Howlett. "Early Phytocannabinoid Chemistry to Endocannabinoids and Beyond." *Nature Reviews Neuroscience* 15, no. 11 (2014): 757–764.

Parsons, Loren H, and Yasmin L Hurd. "Endocannabinoid Signalling in Reward and Addiction." *Nature Reviews Neuroscience* 16 (2015): 579–594.

Whiting, P. F., R. F. Wolff, S. Deshpande, M. Di Nisio, S. Duffy, A. V. Hernandez, J. C. Keurentjes, et al. "Cannabinoids for Medical Use: A Systematic Review and Meta-Analysis." *JAMA* 313 (2015): 2456–2473.

Note: Page numbers in italics indicate figures; page numbers in bold indicate glossary entries.

LINDA A. PARKER is University Faculty Emerita and former Canada Research Chair in Behavioral Neuroscience at the University of Guelph in Ontario, Canada. Her innovative scientific approaches revealing the critical role played by the endocannabinoid system in the regulation of nausea, anxiety, and addiction were recognized by the Lifetime Achievement Award from the International Cannabinoid Research Society (ICRS). She is the author of *Cannabinoids and the Brain* (2017) and a coauthor of *CBD: What Does the Science Say?* (2022), both published by the MIT Press.

Publisher contact:
The MIT Press
Massachusetts Institute of Technology
77 Massachusetts Avenue, Cambridge, MA 02139
mitpress.mit.edu

EU Authorised Representative:
Easy Access System Europe, Mustamäe tee 50,
10621 Tallinn, Estonia
gpsr.requests@easproject.com

Printed by Integrated Books International,
United States of America